Metasurfaces: Physics and Applications

Metasurfaces: Physics and Applications

Special Issue Editors

Sergey I. Bozhevolnyi
Patrice Genevet
Fei Ding

MDPI • Basel • Beijing • Wuhan • Barcelona • Belgrade

MDPI

Special Issue Editors

Sergey I. Bozhevolnyi
University of Southern Denmark
Denmark

Patrice Genevet
Centre de Recherche sur l'Hétéro-Epitaxie et ses Application,
CNRS
France

Fei Ding
University of Southern Denmark
Denmark

Editorial Office
MDPI
St. Alban-Anlage 66
4052 Basel, Switzerland

This is a reprint of articles from the Special Issue published online in the open access journal *Applied Sciences* (ISSN 2076-3417) from 2017 to 2018 (available at: https://www.mdpi.com/journal/applsci/special_issues/metasurfaces)

For citation purposes, cite each article independently as indicated on the article page online and as indicated below:

LastName, A.A.; LastName, B.B.; LastName, C.C. Article Title. *Journal Name* **Year**, *Article Number, Page Range.*

ISBN 978-3-03897-344-7 (Pbk)
ISBN 978-3-03897-345-4 (PDF)

Contents

About the Special Issue Editors

Sergey I. Bozhevolnyi, Professor and Head of Centre for Nano Optics at the University of Southern Denmark in Denmar. Sergey I Bozhevolnyi received his M.Sc. and Ph.D. degrees in Quantum Electronics from Moscow Institute of Physics and Technology (Russia) in 1978 and 1981, respectively. He was conferred the Dr. Sci. degree from Aarhus University, Aarhus, Denmark, in 1998 for his research on subwavelength light confinement. He initiated experimental research in near-field optics at the Institute of Physics, Aalborg University (Denmark), in 1991, where he was Professor in 2003–2009 before moving to SDU. During 2001–2004, he was also the Chief Technical Officer (CTO) of Micro Managed Photons A/S set up to commercialize plasmonic waveguides. He has authored or co-authored more than 450 articles in peer-reviewed international journals, holds 10 patents, and has written 14 book chapters. He has received more than 16800 cites registered by WoS and has an h-index of 59. He has given more than 90 invited talks at international conferences and research seminars. He is currently a Professor and Head of Centre for Nano Optics at the University of Southern Denmark in Denmark. His current research interests include linear and nonlinear nano-optics and plasmonics, including plasmonic interconnects and metasurfaces. He is a Fellow of Optical Society of America (2007) and a member of Danish Academy of Natural Sciences (2010).

Patrice Genevet, CNRS researcher. Dr. Genevet obtained his Ph.D. from the University Côte d'Azur in 2009. He joined the Capasso group at Harvard University in collaboration with Prof. Marlan O. Scully to work on plasmonics and metamaterials. After two years, he became research associate at Harvard University. For three years, he developed the concept of metasurfaces and planar optics. In 2014, he was appointed as a senior research scientist in A*STAR, Singapore. In 2015, he received the ERC stg and joined CNRS to work on semiconductor metasurfaces. He has published 55 papers, including several Nature family journals, Science, Nano Letters, and PRL. He has co-authored five chapters and holds three US patents. He has been invited to more than 30 talks in international conferences, including three plenary presentations at the "Optical MEMS and Nanophotonics Conference" in Banff, at Physics of Quantum Electronics (PQE) 2017, Snowbird USA, and for NANOP 2017 in Barcelona.

Fei Ding, postdoc of Centre for Nano Optics at the University of Southern Denmark in Denmark. Fei Ding received his B.S. and Ph.D. degrees in Optical Engineering from Zhejiang University in 2010 and 2015, respectively. He is currently a postdoctoral fellow at the Centre for Nano Optics at the University of Southern Denmark in Denmark. His current research interests are in the areas of nanophotonics, applied electromagnetics, metasurfaces, and plasmonics, and with a particular focus on innovative and extreme aspects of wave interaction with engineered materials and nanostructures. He has authored or co-authored more than 25 scientific contributions published in peer-reviewed journal papers and peer-reviewed conference proceedings, such as Reports on Progress in Physics, Light: Science & Applications, and ACS Nano, among which there are six ESI highly cited papers and one hot paper, receiving more than 1200 citations on Google Scholar. He is serving as a reviewer for several journals and he is currently the guest-editor of a Special Issue on metasurface for the journal of Applied Sciences. He was awarded the PIERS 2018 Toyama Yong Scientist Award in 2018.

applied
sciences

MDPI

Editorial

Special Issue on "Metasurfaces: Physics and Applications"

Fei Ding [1,*], Patrice Genevet [2] and Sergey I. Bozhevolnyi [1,*]

[1] SDU Nano Optics, University of Southern Denmark, Campusvej 55, DK-5230 Odense, Denmark
[2] Centre de Recherche sur l'Hétéro-Epitaxie et ses Application, CNRS, Rue Bernard Gregory, Sophia-Antipolis, 06560 Valbonne, France; patrice.genevet@crhea.cnrs.fr
* Correspondence: feid@mci.sdu.dk (F.D.); seib@mci.sdu.dk (S.I.B.)

Received: 18 September 2018; Accepted: 20 September 2018; Published: 24 September 2018

Metasurfaces, the two-dimensional analog of metamaterials, have been attracting progressively increasing attention in recent years due to their planar configurations and thus ease of fabrication while enabling unprecedented control in optical fields [1–4]. The phase, amplitude, polarization, helicity, and even angular momentum of the reflected or transmitted optical fields can be controlled at will by tailoring optically thin planar arrays of resonant subwavelength elements arranged in a periodic or aperiodic manner. As a result, numerous applications and fascinating devices have been realized by designed planar metasurfaces, including beam deflectors [5–9], wave plates [10–13], flat lenses [14–20], holograms [21–25], surface wave couplers [26–30], and freeform metasurfaces [31–33].

This special issue is launched to provide a possibility for researchers in the area of metasurfaces to highlight the most recent exciting developments and discuss different metasurface configurations in depth, so as to further promote practical applications of metasurfaces. There are 12 papers selected for this special issue, representing fascinating progress and potential applications in the area of metasurfaces. This collection includes three review papers in total, which focus on a few specific branches of metasurface-based applications [34–36]. Lei Zhou and co-workers present a concise review on the development of multifunctional metasurfaces based on merging concept and anisotropic single-structure meta-atoms [34]. This is a timely overview article, since integrating multiple diversified functionalities into a single and ultra-compact device has become an emerging research area in photonics. The second review paper authored by Wei E.I. Sha and co-workers comprehensively discusses the recent progress in geometric-phase-based metasurfaces for orbital angular momentum (OAM) generation and detection [35]. The last review paper from Bozhevolnyi's group focuses on the fundamentals and recent developments within metasurface-based polarimeters, which can detect the polarization state of an incident beam in one shot with a compact single device [36]. Regarding the other nine research papers, the following metasurface-based application areas are specifically addressed:

Metasurface-based microwave antennas: This special issue contains a series of works on metasurface-based antennas operating in the microwave range, which is an important application of metasurfaces. Long Li and co-workers have utilized well-designed metasurfaces to replace the conventional bulk antennas and demonstrated coherent computational imaging [37], high-order harmonic suppression [38], and electromagnetic power harvesting [39]. Additionally, novel metasurface-based antennas have been proposed with improved characteristics, such as the crossbar fractal microstrip [40] and elliptical patch with cross-shaped aperture [41].

Coding metasurface for beam-steering: This special issue includes two excellent examples of microwave coding metasurfaces, of which one is devoted to wide-angle beam-steering based on 1-bit digital reconfigurable reflective metasurfaces [42], and the other one reports the broadband radar cross-section reduction with linear polarization conversion metasurfaces [43].

Metasurface-based spectrometer: One paper presents a theoretical investigation of an off-axis metalens-based spectrometer by addressing the influences of structural parameters on the effective spectral range and spectral resolution [44]. This study outlines an important way to design and integrate planar metasurface-based spectrometers for various practical applications.

Metasurface-based waveguide: Vladimir P. Drachev and co-workers have demonstrated a metasurface-based waveguide composed of magnetic gratings with effective strips, where anisotropy in the effective parameters is introduced, providing thereby an additional flexibility to control the polarization- and angular-dependent optical response [45].

In summary, this special issue contains a series of excellent research work on metasurfaces, covering a wide area of application-oriented meta-devices. This collection of 12 papers is highly recommended and believed to benefit readers in various aspects.

Acknowledgments: We would like to thank all authors, the many dedicated referees, the editor team of *Applied Sciences*, and especially Ryan Pei (Assistant Managing Editor) for their valuable contributions, making this special issue a success.

Conflicts of Interest: The authors declare no conflicts of interest.

References

1. Yu, N.; Capasso, F. Flat optics with designer metasurfaces. *Nat. Mater.* **2014**, *13*, 139–150. [CrossRef] [PubMed]
2. Glybovski, S.B.; Tretyakov, S.A.; Belov, P.A.; Kivshar, Y.S.; Simovski, C.R. Metasurfaces: From microwaves to visible. *Phys. Rep.* **2016**, *634*, 1–72. [CrossRef]
3. Genevet, P.; Capasso, F.; Aieta, F.; Khorasaninejad, M.; Devlin, R. Recent advances in planar optics: From plasmonic to dielectric metasurfaces. *Optica* **2017**, *4*, 139–152. [CrossRef]
4. Ding, F.; Pors, A.; Bozhevolnyi, S.I. Gradient metasurfaces: A review of fundamentals and applications. *Rep. Prog. Phys.* **2018**, *81*, 026401. [CrossRef] [PubMed]
5. Yu, N.; Genevet, P.; Kats, M.A.; Aieta, F.; Tetienne, J.P.; Capasso, F.; Gaburro, Z. Light Propagation with Phase Discontinuities: Generalized Laws of Reflection and Refraction. *Science* **2011**, *334*, 333–337. [CrossRef] [PubMed]
6. Ni, X.; Emani, N.K.; Kildishev, A.V.; Boltasseva, A.; Shalaev, V.M. Broadband Light Bending with Plasmonic Nanoantennas. *Science* **2012**, *335*, 427. [CrossRef] [PubMed]
7. Sun, S.; Yang, K.Y.; Wang, C.M.; Juan, T.K.; Chen, W.T.; Liao, C.Y.; He, Q.; Xiao, S.; Kung, W.T.; Guo, G.Y.; et al. High-Efficiency Broadband Anomalous Reflection by Gradient Meta-Surfaces. *Nano Lett.* **2012**, *12*, 6223–6229. [CrossRef] [PubMed]
8. Pors, A.; Albrektsen, O.; Radko, I.P.; Bozhevolnyi, S.I. Gap plasmon-based metasurfaces for total control of reflected light. *Sci. Rep.* **2013**, *3*, 2155. [CrossRef] [PubMed]
9. Pors, A.; Ding, F.; Chen, Y.; Radko, I.P.; Bozhevolnyi, S.I. Random-phase metasurfaces at optical wavelengths. *Sci. Rep.* **2016**, *6*, 28448. [CrossRef] [PubMed]
10. Yu, N.; Aieta, F.; Genevet, P.; Kats, M.A.; Gaburro, Z.; Capasso, F. A Broadband, Background-Free Quarter-Wave Plate Based on Plasmonic Metasurfaces. *Nano Lett.* **2012**, *12*, 6328–6333. [CrossRef] [PubMed]
11. Pors, A.; Nielsen, M.G.; Bozhevolnyi, S.I. Broadband plasmonic half-wave plates in reflection. *Opt. Lett.* **2013**, *38*, 513–515. [CrossRef] [PubMed]
12. Yang, Y.; Wang, W.; Moitra, P.; Kravchenko, I.I.; Briggs, D.P.; Valentine, J. Dielectric Meta-Reflectarray for Broadband Linear Polarization Conversion and Optical Vortex Generation. *Nano Lett.* **2014**, *14*, 1394–1399. [CrossRef] [PubMed]
13. Ding, F.; Wang, Z.; He, S.; Shalaev, V.M.; Kildishev, A.V. Broadband High-Efficiency Half-Wave Plate: A Supercell-Based Plasmonic Metasurface Approach. *ACS Nano* **2015**, *9*, 4111–4119. [CrossRef] [PubMed]
14. Aieta, F.; Genevet, P.; Kats, M.A.; Yu, N.; Blanchard, R.; Gaburro, Z.; Capasso, F. Aberration-Free Ultrathin Flat Lenses and Axicons at Telecom Wavelengths Based on Plasmonic Metasurfaces. *Nano Lett.* **2012**, *12*, 4932–4936. [CrossRef] [PubMed]
15. Ni, X.; Ishii, S.; Kildishev, A.V.; Shalaev, V.M. Ultra-thin, planar, Babinet-inverted plasmonic metalenses. *Light Sci. Appl.* **2013**, *2*, e72. [CrossRef]
16. Pors, A.; Nielsen, M.G.; Eriksen, R.L.; Bozhevolnyi, S.I. Broadband Focusing Flat Mirrors Based on Plasmonic Gradient Metasurfaces. *Nano Lett.* **2013**, *13*, 829–834. [CrossRef] [PubMed]

17. Arbabi, A.; Horie, Y.; Bagheri, M.; Faraon, A. Dielectric metasurfaces for complete control of phase and polarization with subwavelength spatial resolution and high transmission. *Nat. Nanotechnol.* **2015**, *10*, 937–943. [CrossRef] [PubMed]

18. Khorasaninejad, M.; Chen, W.T.; Devlin, R.C.; Oh, J.; Zhu, A.Y.; Capasso, F. Metalenses at visible wavelengths: Diffraction-limited focusing and subwavelength resolution imaging. *Science* **2016**, *352*, 1190–1194. [CrossRef] [PubMed]

19. Arbabi, A.; Arbabi, E.; Kamali, S.M.; Horie, Y.; Han, S.; Faraon, A. Miniature optical planar camera based on a wide-angle metasurface doublet corrected for monochromatic aberrations. *Nat. Commun.* **2016**, *7*, 13682. [CrossRef] [PubMed]

20. Wang, S.; Wu, P.C.; Su, V.C.; Lai, Y.C.; Chu, C.H.; Chen, J.W.; Lu, S.H.; Chen, J.; Xu, B.; Kuan, C.H.; et al. Broadband achromatic optical metasurface devices. *Nat. Commun.* **2017**, *8*, 187. [CrossRef] [PubMed]

21. Ni, X.; Kildishev, A.V.; Shalaev, V.M. Metasurface holograms for visible light. *Nat. Commun.* **2013**, *4*, 2807. [CrossRef]

22. Chen, W.T.; Yang, K.Y.; Wang, C.M.; Huang, Y.W.; Sun, G.; Chiang, I.D.; Liao, C.Y.; Hsu, W.L.; Lin, H.T.; Sun, S.; et al. High-efficiency broadband meta-hologram with polarization-controlled dual images. *Nano Lett.* **2013**, *14*, 225–230. [CrossRef] [PubMed]

23. Huang, L.; Chen, X.; Mühlenbernd, H.; Zhang, H.; Chen, S.; Bai, B.; Tan, Q.; Jin, G.; Cheah, K.W.; Qiu, C.W.; et al. Three-dimensional optical holography using a plasmonic metasurface. *Nat. Commun.* **2013**, *4*, 2808. [CrossRef]

24. Zheng, G.; Mühlenbernd, H.; Kenney, M.; Li, G.; Zentgraf, T.; Zhang, S. Metasurface holograms reaching 80% efficiency. *Nat. Nanotechnol.* **2015**, *10*, 308–312. [CrossRef] [PubMed]

25. Genevet, P.; Capasso, F. Holographic optical metasurfaces: A review of current progress. *Rep. Prog. Phys.* **2015**, *78*, 024401. [CrossRef] [PubMed]

26. Sun, S.; He, Q.; Xiao, S.; Xu, Q.; Li, X.; Zhou, L. Gradient-index meta-surfaces as a bridge linking propagating waves and surface waves. *Nat. Mater.* **2012**, *11*, 426–431. [CrossRef] [PubMed]

27. Lin, J.; Mueller, J.B.; Wang, Q.; Yuan, G.; Antoniou, N.; Yuan, X.C.; Capasso, F. Polarization-controlled tunable directional coupling of surface plasmon polaritons. *Science* **2013**, *340*, 331–334. [CrossRef] [PubMed]

28. Huang, L.; Chen, X.; Bai, B.; Tan, Q.; Jin, G.; Zentgraf, T.; Zhang, S. Helicity dependent directional surface plasmon polariton excitation using a metasurface with interfacial phase discontinuity. *Light Sci. Appl.* **2013**, *2*, e70. [CrossRef]

29. Pors, A.; Nielsen, M.G.; Bernardin, T.; Weeber, J.C.; Bozhevolnyi, S.I. Efficient unidirectional polarization-controlled excitation of surface plasmon polaritons. *Light Sci. Appl.* **2014**, *3*, e197. [CrossRef]

30. Ding, F.; Deshpande, R.; Bozhevolnyi, S.I. Bifunctional gap-plasmon metasurfaces for visible light: Polarization-controlled unidirectional surface plasmon excitation and beam steering at normal incidence. *Light Sci. Appl.* **2018**, *7*, 17178. [CrossRef]

31. Teo, J.T.H.; Wong, L.J.; Molardi, C.; Genevet, P. Controlling electromagnetic fields at boundaries of arbitrary geometries. *Phys. Rev. A* **2016**, *94*, 023820. [CrossRef]

32. Kamali, S.M.; Arbabi, A.; Arbabi, E.; Horie, Y.; Faraon, A. Decoupling optical function and geometrical form using conformal flexible dielectric metasurfaces. *Nat. Commun.* **2016**, *7*, 11618. [CrossRef] [PubMed]

33. Wu, K.; Coquet, P.; Wang, Q.J.; Genevet, P. Modelling of Free-form Conformal Metasurfaces. *Nat. Commun.* **2018**, *9*, 3494. [CrossRef] [PubMed]

34. Tang, S.; Cai, T.; Xu, H.; He, Q.; Sun, S.; Zhou, L. Multifunctional Metasurfaces Based on the "Merging" Concept and Anisotropic Single-Structure Meta-Atoms. *Appl. Sci.* **2018**, *8*, 555. [CrossRef]

35. Chen, M.; Jiang, L.; Sha, W. Orbital Angular Momentum Generation and Detection by Geometric-Phase Based Metasurfaces. *Appl. Sci.* **2018**, *8*, 362. [CrossRef]

36. Ding, F.; Chen, Y.; Bozhevolnyi, S.I. Metasurface-Based Polarimeters. *Appl. Sci.* **2018**, *8*, 594. [CrossRef]

37. Kou, N.; Li, L.; Tian, S.; Li, Y. Measurement Matrix Analysis and Radiation Improvement of a Metamaterial Aperture Antenna for Coherent Computational Imaging. *Appl. Sci.* **2017**, *7*, 933. [CrossRef]

38. Kou, N.; Liu, H.; Li, L. A Transplantable Frequency Selective Metasurface for High-Order Harmonic Suppression. *Appl. Sci.* **2017**, *7*, 1240. [CrossRef]

39. Zhang, X.; Liu, H.; Li, L. Electromagnetic Power Harvester Using Wide-Angle and Polarization-Insensitive Metasurfaces. *Appl. Sci.* **2018**, *8*, 497. [CrossRef]

40. Kubacki, R.; Czyżewski, M.; Laskowski, D. Minkowski Island and Crossbar Fractal Microstrip Antennas for Broadband Applications. *Appl. Sci.* **2018**, *8*, 334. [CrossRef]
41. Tellechea, A.; Ederra, I.; Gonzalo, R.; Iriarte, J.C. Dispersion Properties of an Elliptical Patch with Cross-Shaped Aperture for Synchronized Propagation of Transverse Magnetic and Electric Surface Waves. *Appl. Sci.* **2018**, *8*, 472. [CrossRef]
42. Tian, S.; Liu, H.; Li, L. Design of 1-Bit Digital Reconfigurable Reflective Metasurface for Beam-Scanning. *Appl. Sci.* **2017**, *8*, 882. [CrossRef]
43. Yang, J.; Cheng, Y.; Qi, D.; Gong, R. Study of Energy Scattering Relation and RCS Reduction Characteristic of Matrix-Type Coding Metasurface. *Appl. Sci.* **2018**, *8*, 1231. [CrossRef]
44. Zhou, Y.; Chen, R.; Ma, Y. Characteristic Analysis of Compact Spectrometer Based on Off-Axis Meta-Lens. *Appl. Sci.* **2018**, *8*, 321. [CrossRef]
45. Roccapriore, K.M.; Lyvers, D.P.; Brown, D.P.; Poutrina, E.; Urbas, A.M.; Germer, T.A.; Drachev, V.P. Waveguide Coupling via Magnetic Gratings with Effective Strips. *Appl. Sci.* **2018**, *8*, 6177. [CrossRef]

applied
sciences

MDPI

Review

Multifunctional Metasurfaces Based on the "Merging" Concept and Anisotropic Single-Structure Meta-Atoms

Shiwei Tang [1,*], Tong Cai [2], He-Xiu Xu [2], Qiong He [3], Shulin Sun [4] and Lei Zhou [3,*]

1   Department of Physics, Faculty of Science, Ningbo University, Ningbo 315211, China
2   Air and Missile Defend College, Air force Engineering University, Xi'an 710051, China;
    caitong326@sina.cn (T.C.); hxxuellen@gmail.com (H.-X.X.)
3   State Key Laboratory of Surface Physics, Key Laboratory of Micro and Nano Photonic Structures (Ministry of Education), Collaborative Innovation Center of Advanced Microstructures, and Physics Department of Fudan University, Shanghai 200433, China; qionghe@fudan.edu.cn
4   Shanghai Engineering Research Center of Ultra-Precision Optical Manufacturing, Green Photonics and Department of Optical Science and Engineering, Fudan University, Shanghai 200433, China;
    sls@fudan.edu.cn
*   Correspondence: tsw@fudan.edu.cn (S.T.); phzhou@fudan.edu.cn (L.Z.);
    Tel.: +86-180-067-06998 (S.T.); +86-021-5566-5236 (L.Z.)

Received: 7 March 2018; Accepted: 2 April 2018; Published: 4 April 2018

Abstract: Metasurfaces offer great opportunities to control electromagnetic (EM) waves, attracting intensive attention in science and engineering communities. Recently, many efforts were devoted to multifunctional metasurfaces integrating different functionalities into single flat devices. In this article, we present a concise review on the development of multifunctional metasurfaces, focusing on the design strategies proposed and functional devices realized. We first briefly review the early efforts on designing such systems, which simply combine multiple meta-structures with distinct functionalities to form multifunctional devices. To overcome the low-efficiency and functionality cross-talking issues, a new strategy was proposed, in which the meta-atoms are carefully designed single structures exhibiting polarization-controlled transmission/reflection amplitude/phase responses. Based on this new scheme, various types of multifunctional devices were realized in different frequency domains, which exhibit diversified functionalities (e.g., focusing, deflection, surface wave conversion, multi-beam emissions, etc.), for both pure-reflection and pure-transmission geometries or even in the full EM space. We conclude this review by presenting our perspectives on this fast-developing new sub-field, hoping to stimulate new research outputs that are useful in future applications.

Keywords: metasurface; multifunctional device; metamaterial; meta-atom

1. Introduction

Facing the increasing demands on data-storage capacity and information processing speed in modern science and technology, electromagnetic (EM) integration plays a more and more important role, which has intrigued intensive attention with remarkable applications. An ultimate goal pursued by scientists and engineers along this development is to make miniaturized devices as small as possible, yet equipped with powerful functionalities as many as possible. However, available efforts based on conventional materials suffer from the issues of device thickness, low efficiency, and restricted functionalities, caused by the fact that natural materials only exhibit electric responses with small variation range of permittivity ε and, thus, they only have limited capabilities on manipulating EM waves.

Metamaterials (MTMs) [1,2], consisting of deep-subwavelength-sized EM microstructures (e.g., meta-atoms) arranged in periodic or non-periodic orders, have drawn much attention recently. Through tailoring the microstructures of meta-atoms, MTMs can, in principle, exhibit arbitrary values of permittivity ε and permeability μ, which offers MTMs extraordinarily strong capabilities to control EM waves. Many fascinating wave-manipulation effects have been demonstrated based on MTMs, such as negative refraction [3,4], super-resolution imaging [5,6], cloaking [7–9], polarization-control [10–13], perfect light absorption [14], and transparency [15,16], and unusual wave-control effects realized by zero-index MTMs [17,18]. Attempts have also been made to achieve multifunctional EM devices based on MTMs. However, the realized devices typically exhibit bulky sizes and low efficiencies, since MTMs are three-dimensional (3D) materials composed by resonant metallic structures which can easily absorb EM waves. Moreover, such 3D devices require complex fabrication processes, adding more disadvantages for EM integration [19,20].

Metasurfaces, ultrathin MTM layers constructed by planar meta-atoms of pre-determined EM responses arranged in specific two-dimensional (2D) orders, can largely overcome the difficulties faced by MTMs. Tuning the EM responses of meta-atoms to realize certain transmission/reflection phase distributions on the metasurfaces, one can use these ultra-thin devices to efficiently reshape the wave-fronts of incident EM beams based on Huygens' principle, achieving unusual effects, including anomalous beam bending based on generalized Snell's law [21–29], propagating wave to surface waves conversion [30–32], polarization-control [33–39], focusing [40–42], holograms [43–45], flat-lens imaging [46–50], tunable devices [51–53], and photonic spin-Hall effect [54–56], etc. Typically, these devices are flat, much thinner than the wavelength, and exhibit much higher efficiencies than their bulky MTM counterparts, all being highly favorable for integration-optics applications. These attractive properties make metasurfaces the best candidates to construct multifunctional EM devices. Indeed, many efforts have recently been devoted to designing multifunctional optical devices based on metasurfaces [57–87], typically using polarization or frequency of the incident light as an external knob to control the functionality exhibited by the devices. The proposed/demonstrated devices are usually equipped with functionalities combining two or more from those demonstrated before on single-function metasurfaces, such as beam-bending, focusing, hologram, surface-wave conversion, and directive beaming, etc.

In this paper, we present a concise review on this fast developing sub-field, focusing on the working principles and practical realizations of multi-functional metasurface-based meta-devices. We particularly emphasize the importance of designing appropriate meta-atoms in this field, since the remaining tasks are merely engineering optimizations after high-performances meta-atoms are found. This review is organized as follows. We first briefly summarize in Section 2 a class of multifunctional meta-devices based on the "merging" concept. Having understood the merits and disadvantages of the "merging" scheme, we then introduce another class of multifunctional metasurfaces based on *single-structure* meta-atoms exhibiting polarization-controlled transmission/reflection phase responses. The realized meta-devices include those exhibiting *similar* wave-control functionalities for two polarizations (Section 3.1), those integrating *different* functionalities with very high efficiencies in reflection geometry (Section 3.2) and in transmission geometry (Section 3.3), and those utilizing the full EM space to control EM waves with different capabilities (Section 3.4). We finally conclude this review and present our perspectives in the last section.

2. Multifunctional Meta-Devices Based on the "Merging" Concept

A simple scheme developed in early years utilized the so-called "merged" meta-structures to design multifunctional metasurfaces. In such a scheme, people first design individual metasurfaces exhibiting their own functions and then construct a multifunctional device simply through merging the two structures together. Below we present several examples to illustrate how the scheme works.

Figure 1a presents an optical bifunctional metasurface that can realize a hologram image or a vortex beam, depending on the helicity of excitation light [74]. To achieve their end, the authors first design two individual metasurfaces (both utilizing the metal-bar structure as basic meta-atoms) which can realize one of the needed functionalities when they are shined by incident light taking circular polarizations (CP) with different helicities (see Figure 1a). The desired phase profiles on two metasurfaces are created by the Pancharatnam-Berry (PB) principle [54,88,89] through rotating the metallic bars at different positions by appropriate angles. Since the two metasurfaces exhibit identical periodic structures and there are enough open spaces between metallic bars, the authors then merge two metasurfaces together to obtain the final design in which all metallic bars do not touch with each other. Such a device was finally fabricated out and experimentally characterized, showing nice bifunctional performances (Figure 1a). However, the working efficiency of the device is quite low, which is found to be around 9% [74].

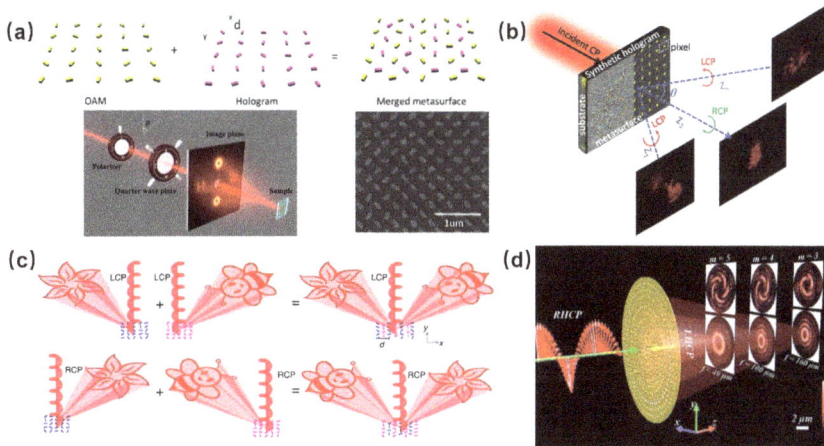

Figure 1. Multifunctional devices designed with merged structures. (**a**) Design strategy, sample picture, and experimental characterizations of a multifunctional metasurface than can generate holographic images or a vortex beam depending on the helicity of incident circularly polarized light. Reproduced from Ref. [74] with permission. (**b**) A metasurface that can reconstruct different images to different polarization channels. Reproduced from [64] with permission. (**c**) A metasurface that can generate multiple hologram images as shined by circularly polarized light with different helicity. Reproduced from [63] with permission. (**d**) A metasurface that can generate optical vortices with distinct topological charges at different longitudinal focal planes. Reproduced from [58] with permission. LCP, left CP; RCP, right CP; OAM, orbital angular momentum; LHCP, left-handed circularly polarized ; RHCP, right-handed circularly polarized.

Such a "merging" concept has been straightforwardly applied to realize many other multifunctional meta-devices [58,61–65]. Figure 1c presents a reflective bifunctional meta-device which can form a holographic image of a 'flower' or a 'bee' as the helicity of incident light changes from left CP (LCP) to right CP (RCP) [63]. In Figure 1b, a new type of meta-hologram device is introduced, which can exhibit various hologram patterns depending on the helicity the incident light [64]. Figure 1d shows a meta-device which can control the polarization state (i.e., spin angular momentum) and the position of the focal planes by manipulating the helicity of the incident light. [58].

In reviewing these meta-devices based on the "merging" concept, we find that the proposed design strategy is physically transparent and easy to implement. However, to make the "merging" process work, the adopted meta-atoms must be very simple structures (say, metal bar) to avoid metallic overlapping. Unfortunately, these meta-atoms typically do not satisfy the 100%-efficiency criterion established for PB metasurfaces [54] and, thus, one type of meta-atoms can generate background noise in addition to the desired functionalities. As a result, such meta-devices typically suffer from issues of low operating efficiencies and functionality cross-talking, except [63] where the issue was partially solved by seeking a high-efficiency PB meta-atom in the reflection geometry.

3. Multifunctional Metasurfaces Based on Anisotropic Single-Structure Meta-Atoms for Two Polarizations

Having understood the key issues in the "merging" concept, people propose a new strategy to design multifunctional metasurfaces, i.e., using *single* structures as the basic meta-atoms to design the final device instead of merging two separately determined ones. In what follows, we first discuss the basic concept, and then introduce several meta-devices realized with such a strategy, classified into four sub-categories as detailed below.

Suppose that the adopted meta-atoms possess mirror symmetries, we can then describe the EM response for a meta-atom located at a position (x, y) on a metasurface by two diagonal Jones's matrices:

$$\mathbf{R} = \begin{pmatrix} r_{xx}(x,y) & 0 \\ 0 & r_{yy}(x,y) \end{pmatrix} \tag{1}$$

and:

$$\mathbf{T} = \begin{pmatrix} t_{xx}(x,y) & 0 \\ 0 & t_{yy}(x,y) \end{pmatrix} \tag{2}$$

where $r_{xx}, r_{yy}, t_{xx}, t_{yy}$ denote the reflection/transmission coefficients of the meat-atom (periodically repeated to form a periodic metasurface), respectively. In order to make the devices exhibit working efficiencies as high as possible, ideally one requires the designed meta-atoms to be either perfectly reflective:

$$\mathbf{T} = 0; \ |r_{ii}| = 1 \tag{3}$$

or perfectly transparent:

$$\mathbf{R} = 0; \ |t_{ii}| = 1. \tag{4}$$

Once such conditions are satisfied meaning that all meta-atoms are of highest working efficiencies, people then adjust the geometrical parameters of each meta-atom to make the whole metasurface exhibit desired polarization-dependent phase profiles (i.e., $\varphi_{xx}(x,y)$ and $\varphi_{yy}(x,y)$) to achieve certain wave-manipulation functionalities. A typical example is schematically shown in Figure 2, for which the wanted bi-functionalities are focusing and beam bending, respectively. To achieve this end, the final phase distributions on the metasurface should be:

$$\varphi_{yy}(x,y) = k_0(\sqrt{F^2 + x^2 + y^2} - F) \tag{5}$$

with F being the focal length and k_0 being the free-space wave-vector, and:

$$\varphi_{xx}(x,y) = C_1 + \xi \cdot x \tag{6}$$

with ξ being the phase gradient responsible for the anomalous reflection angle and C_1 being an a constant. We note that a conversion from propagating wave (PW) to surface wave (SW) can happen if $\xi > k_0$.

We emphasize that the high-efficiency conditions Equations (1) and (2) can only be approximately satisfied since losses inevitably exist, especially at high frequencies. Moreover, the achieved functionalities are not limited to the two described above. If different wave-manipulation functionalities are desired, one may easily replace Equations (5) and (6) by the phase distributions related to those functionalities.

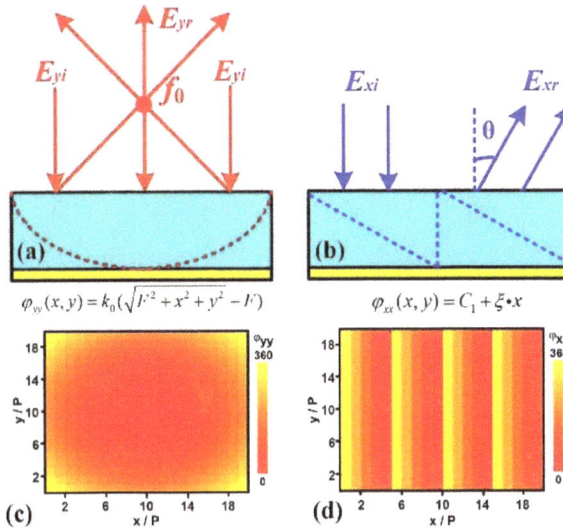

Figure 2. Schematics of a bi-functional metasurface (blue) with metal ground plane (yellow) which can achieve (**a**) focusing functionality for *y*-polarized incident light and (**b**) anomalous-reflection functionality for *x*-polarized incident light. The phase distributions of the metasurface are depicted in (**c**) for *y*-polarization excitation and in (**d**) for *x*-polarization excitation. Reproduced from [90] with permission.

3.1. Multifunctional Metasurfaces Exhibiting Similar Functionalities

Historically, multifunctional metasurfaces designed by single-structure meta-atoms were first realized in reflection geometry, because high-efficiency reflective meta-atoms are much easier to find than their transmissive counterparts. A commonly-used reflective meta-atom is the so-called metal-insulator-metal (MIM) structure, which consists of a metallic planar resonator and a continuous metal sheet separated by a dielectric spacer (see Figure 3a). Such an MIM structure was widely used to design high-efficiency reflective metasurfaces at frequencies ranging from microwave to visible [22,23,44,72,79,91], simply because its metallic ground plane can help reflect all incoming waves back (thus, Equation (1) can be easily satisfied) while the phase of reflected wave can undergo a continuous $-180°$ to $180°$ variation as frequency passes through a resonance inside the "meta-atom". Such a resonance is usually called "magnetic resonance" [14] or "gap-plasmon resonance" [22,72], and its properties are carefully analyzed in [91,92]. Therefore, tuning the geometrical parameters of such a meta-atom can efficiently modify its two reflection phases (φ_{xx} and φ_{yy}) (see Figure 3b), which then offers enough freedoms to design multifunctional meta-devices.

Figure 3d depicts a bifunctional meta-device working at telecom wavelengths, which can excite surface plasmon polaritons (SPPs) to flow along two orthogonal directions when the device is shined by external light with polarizations along *x* or *y* direction [72]. The fabricated device consists of 3 × 3 MIM meta-atoms, with geometrical parameters carefully adjusted such that both $\varphi_{xx}(x,y)$ and $\varphi_{yy}(x,y)$ satisfy Equation (6) with $\xi > k_0$. Experimental characterizations reveal that the fabricated device can indeed exhibit the desired bi-functionalities (see Figure 3a). Based the same concept, other bi-functional meta-devices were realized based on MIM meta-atoms with patches replaced by other type of planar resonators, to achieve the polarization-dependent anomalous reflection (beam splitting) (Figure 3c) [79] and polarization-controlled dual-image hologram (Figure 3e) [59]. Two common features of these meta-devices are worth mentioning: the realized functionalities are identical or similar, and the working efficiencies are not very high, partially caused by the metallic losses at optical frequencies.

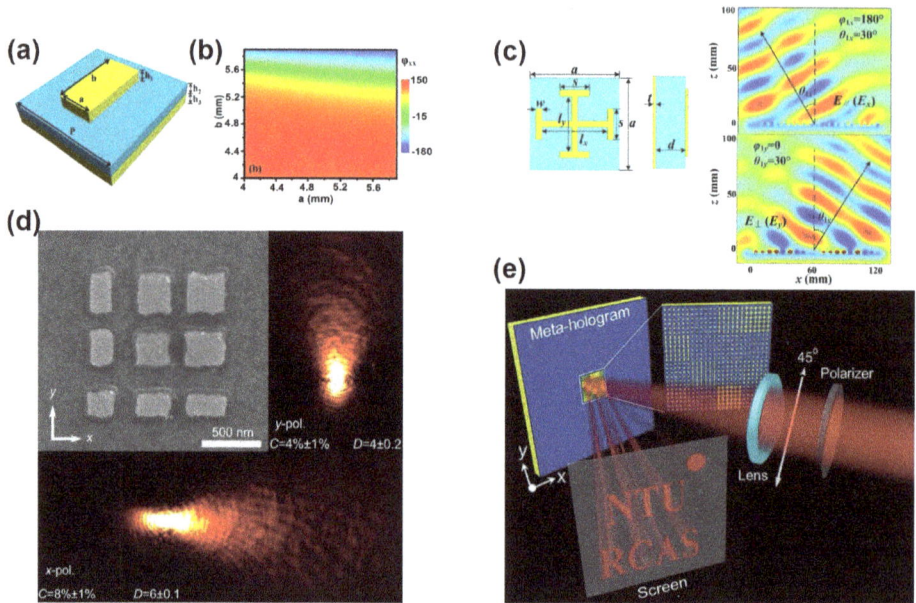

Figure 3. (**a**) Schematics of an MIM meta-atom consisting of a metallic patch resonator and a metal ground plane (yellow) separated by a dielectric spacer (blue). (**b**) Tuning the reflection phase φ_{xx} of a microwave MIM meta-atom by varying a and b. Reproduced from [93] with permission. (**c**) A polarization beam splitter made by bifunctional gradient metasurface constructed with MIM meta-atoms consisting of metal-cross planar resonators and metal ground plane (yellow) separated by a dielectric spacer (blue). Reproduced from [79] with permission. (**d**) Experimental characterizations on a unidirectional polarization-controlled SPP (red) coupler at telecom wavelength. Reproduced from [72] with permission. (**e**) A bi-functional meta-hologram device consisting of metal pattern and metal ground plane (yellow) separated by a dielectric spacer (blue) that can generate different hologram image depending on the incident linear polarization. Reproduced from [59] with permission.

3.2. Multifunctional Reflective Metasurfaces Combining Distinct Functionalities

The proposed design strategy is so general that it can also be used to realize multifunctional metasurfaces exhibiting *distinct* functionalities. For example, the first reflective multifunctional meta-device possessing two distinct functionalities (focusing and PW-to-SW conversion) was demonstrated in [93] in the microwave regime (see Figure 4a,b). Taking meta-atoms the same as shown in Figure 3a, the authors successfully designed a reflective bi-functional metasurface which exhibits two phase distributions satisfying Equations (5) and (6). Shining a *y*-polarized microwave at frequency 9.3 GHz normally onto the sample, the authors measured the $\mathrm{Re}[\vec{E}]$ distributions on two orthogonal planes. Results displayed in Figure 4c indicate clearly that the incoming plane wave has been focused to a focal point. To demonstrate the PW-to-SW conversion effect for the $\vec{E} \parallel \hat{x}$ polarization, the authors purposely fabricated a mushroom structure supporting eigen spoof SPP [30,32] and then put it at the right side of the metasurface to guide out the driven SW (see inset to Figure 4d). Shining the metasurface by an *x*-polarized microwave (9.3 GHz), the authors used a near-field scanning technique to map out the $\mathrm{Re}[E_z]$ field pattern (see Figure 4d). The measured field pattern represents a very well-defined spoof SPP with $k_{spp} = 206.5$ m^{-1}, in good agreement with the theoretically-calculated value ($k_{spp} = 209.4$ m^{-1}). For both functionalities, the experimentally-estimated working efficiencies are higher than 90%, which is not surprising since metallic losses are negligible at microwave frequencies. Such a concept was successfully extended to realize the same type of bifunctional meta-device, but working at optical wavelengths (Figure 4e) [94], which is more close to realistic integration-optics applications.

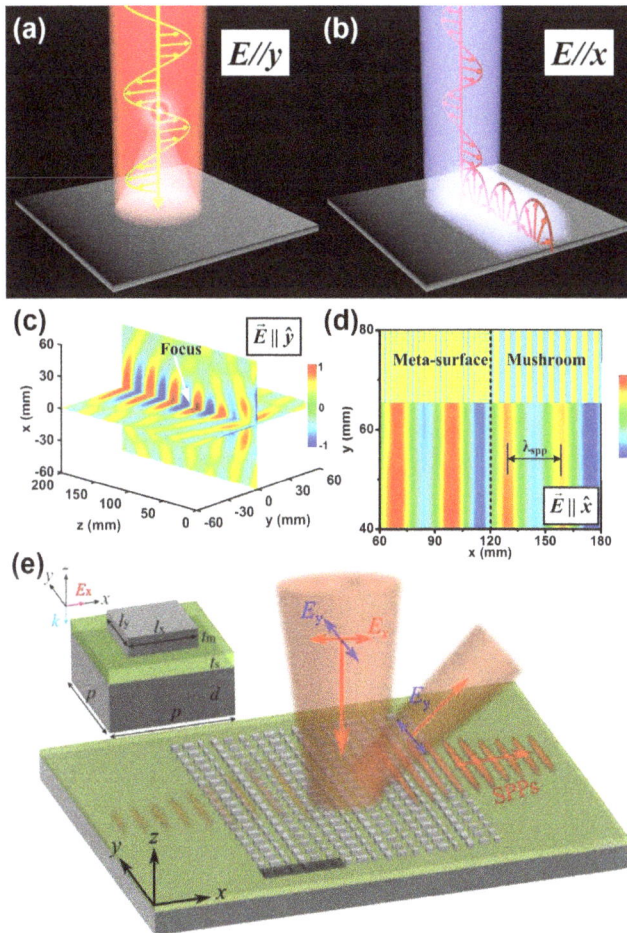

Figure 4. A reflective bifunctional metasurfaces that behaves as (**a**) a focusing lens and (**b**) a PW-to-SW (propagating wave to surface wave) convertor when excited by incident waves with polarizations $E//y$ (red light) and $E//x$ (blue light), respectively. (**c**) Measured Re[$\vec{E}$] distributions on both *xoz* and *yoz* planes as the metasurface is illuminated by a normally incident *y*-polarized plane wave. (**d**) Measured Re[$\vec{E}$] pattern on the *xy*-plane using a monopole antenna placed vertically and 8 mm above the metasurface and the mushroom structure, when the metasurface is illuminated by a normally incident *x*-polarized plane wave. Reproduced from [93] with permission. (**e**) The working principle of similar type of bi-functional metasurface operating for visible light realized in [94]. The top panel shows the schematic of the unit cell consisting of an Ag nanobrick (grey) on top of a spacer (green) and Ag substrate. Reproduced from [94] with permission. SPPs, surface plasmon polaritons.

In designing these devises, people noticed that an arbitrary anisotropic meta-atom might exhibit undesired polarization cross-talking (i.e., varying one parameter of the meta-atom can influence its EM responses for both polarizations). Such an effect hinders the fast designs of meta-devices with complex functionalities, which typically require 2D parameter searching to realize the desired phase profiles. To overcome such cross-talking and to enlarge the working bandwidth, Xu et al. proposed a new type of anisotropic meta-atom [95] which consists of multilayers of planar resonators coupled with a metallic ground plane (Figure 5a). A crucial improvement is to add a wire loop to surround each planar resonator (see Figure 5a), which can significantly degrade the polarization cross-talking due to the screening effect [95].

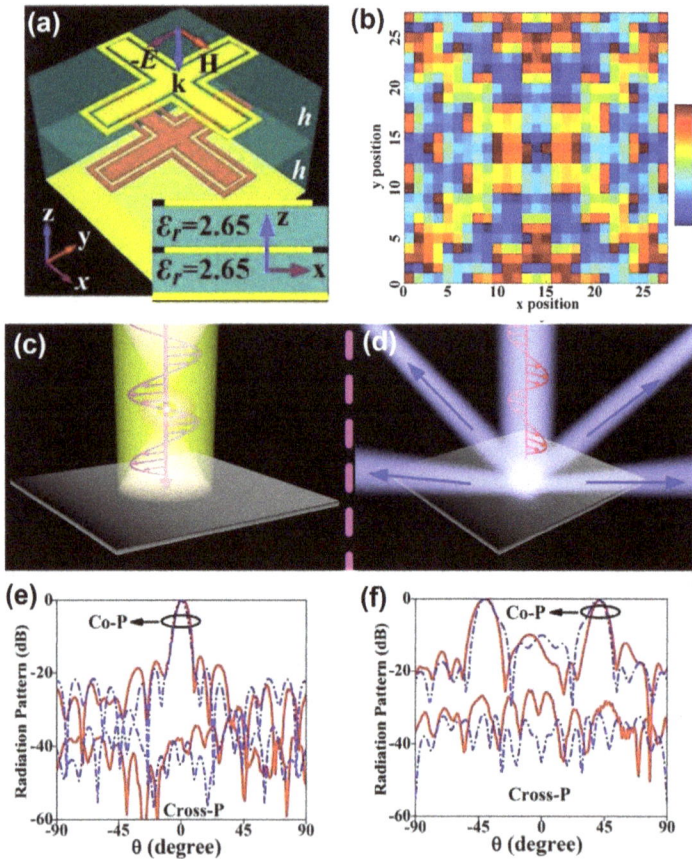

Figure 5. (a) Topology of the dual-layer anisotropic meta-atoms using composite cross bar and cross loop. The meta-atom contains two identical composite metallic resonators and a continuous metal plate (yellow) separated by two dielectric spacers (blue). (b) Phase distribution required on the metasurface to achieve quad-beam emissions. Schematics of a bifunctional metasurface which behaves as (c) a lens or (d) a beam splitter to generate quad large-angle pencil beams when excited by incident waves with polarizations $E//x$ (yellow light) and $E//y$ (blue light), respectively. (e,f) Simulated (red lines) and measured (blue dashed lines) radiation patterns on *x-z* plane for *x* polarization and *y* polarization, demonstrating the bi-functionality possessed by the fabricated device. Reproduced from [95] with permission.

Based on such meta-atoms, the authors successfully realized two bifunctional metasurfaces combining complex wave-control functionalities [95]. One device is shown in Figure 5 which can achieve the functionalities of focusing (Figure 5c) and large-angle multi-beam emissions (Figure 5d) under external excitations with two polarizations. While the phase distribution responsible for the focusing functionality is easy to obtain (Equation (5)), that for another one needs to be carefully determined (see Figure 5b). The structural parameters of all meta-atoms can be quickly determined from two phase profiles, thanks to the low polarization cross talking property of the meta-atoms. Measured radiation patterns (Figure 5e,f) demonstrated clearly the bi-functionalities possessed by the fabricated device. Another device realized in [95] can deflect beam anomalously and achieve small-angle multi-beam emissions for two polarizations. With such a unique meta-atom structure at hand, people can realize other multifunctional meta-devices exhibiting powerful and complex wave-manipulation functionalities.

3.3. Multifunctional Transmissive Metasurfaces Combining Distinct Functionalities

Compared to reflective devices, high-efficiency multifunctional meta-devices in transmission geometry are much more difficult to realize, simply because more channels (transmission and reflection ones) exist in this geometry and thus Equation (2) is difficult to meet. People utilized the concept of Huygens' surface to construct transmission-mode metasurfaces with high efficiencies, but such devices usually require complex 3D non-flat meta-atoms with both electric and magnetic resonators incorporated [24,26]. Due to the complexities in meta-atom design, multifunctional meta-devices realized based on such a scheme are rarely seen, except a transmissive functionality-tunable device recently realized with controllable active elements incorporated [96].

Meanwhile, meta-atoms in multilayer geometry (with deep-subwavelength total thicknesses) are found as alternative candidates to construct high-efficiency transmission-mode metasurfaces. Many functional meta-devices are fabricated using such kind of meta-atoms [27,49], generating interesting effects such as high-efficiency SPP excitations [31] and photonic spin-Hall effect [54]. Straightforwardly, there appear recent efforts on using such meta-atoms to construct transmissive multifunctional meta-devices with very high efficiencies.

Cai et al. proposed to use a four-layer structure, with each layer containing a metallic mesh coupled with a metal patch [93], as the basic meta-atom to construct transmissive bifunctional meta-devices (Figure 6a). Each single layer exhibits perfect EM transmission at a particular frequency due to the interaction between the patch resonator and the opaque mesh. Mutual interactions between different layers can significantly enlarge the transparency window and the transmission-phase variation range (see Figure 6b). With such a high-performance structure, the authors designed and fabricated a transmissive metasurface (see Figure 6a for sample picture) with meta-atoms carefully adjusted to yield transmission phases (φ_{xx} and φ_{yy}) satisfying Equations (5) and (6). Experimental results shown in Figure 6e,f indicate that the device can focus y-polarized PW to a point and can refract x-polarized PW anomalously, illustrating the desired bi-functionality. The achieved working efficiency is 72%, much higher than those of other transmissive metasurfaces [25–27,49].

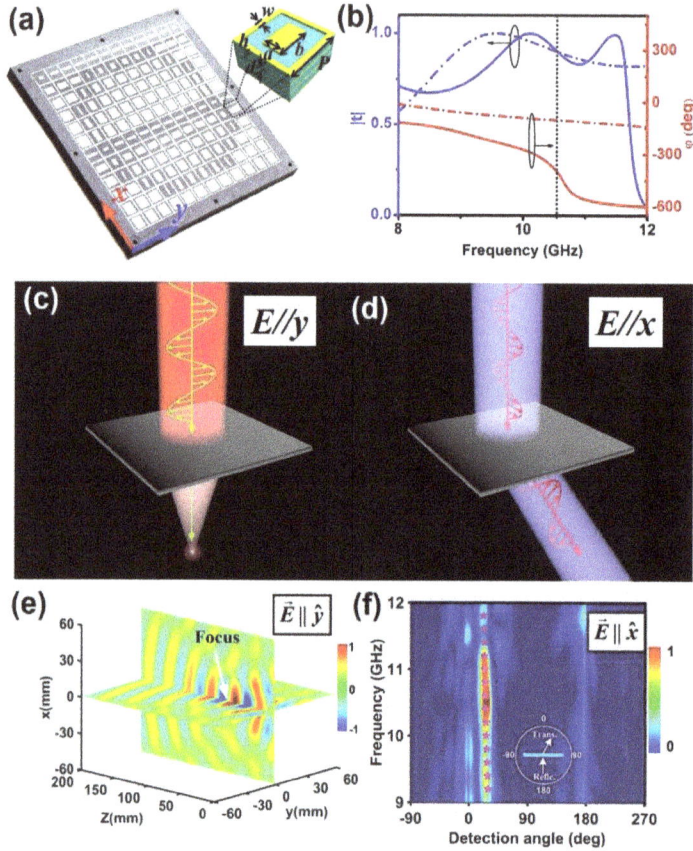

Figure 6. (**a**) Photograph of a microwave transmissive bifunctional metasurface. Inset illustrates a typical meta-atom composed by four metallic layers (yellow) separated by three F4B spacers (blue). (**b**) Transmission amplitude (blue lines) and phase (red lines) for a periodic metasurface constructed by the 4-layer meta-atom, under the excitations of *y*-polarized (solid lines) and *x*-polarized (dotted lines) incident waves, respectively. Schematics and working principles of a transmissive bifunctional metasurfaces, which behaves as (**c**) a focusing lens and (**d**) a beam deflector when excited by incident waves with polarizations $E//y$ and $E//x$, respectively. (**e**) Measured $\mathrm{Re}[\vec{E}]$ distributions on both *xoz* and *yoz* planes as the metasurface is illuminated by a normally-incident *y*-polarized plane wave. (**f**) Measured scattered wave intensity as function of frequency and detection angle when the metasurface is illuminated by *x*-polarized normally incident plane waves. Reproduced from [93] with permission.

3.4. Multifunctional Metasurfaces for Full-Space Manipulation of EM Waves

The multifunctional metasurfaces introduced in the previous sections, whether working in reflection or transmission geometries, leave half of the EM space completely unexplored. It is highly desired to expand the wave-manipulation capabilities of metasurfaces to the full EM space, offering the metasurfaces independently controlled functionalities at their two different sides. In this section, we introduce recent efforts to design metasurfaces that can manipulate the wave fronts of EM waves in the full EM space with very high efficiencies [97].

The key step in this new strategy is to design a collection of meta-atoms which are perfectly transparent or reflective for incident waves polarized along two orthogonal directions. As shown in the inset to Figure 7a, the meta-atom consists of four metallic layers separated by three dielectric spacers. A crucial difference of present meta-atom with a usual transmissive one (Figure 6a) is that here the *x*-orientated metallic stripes of the bottom two layers are *continuous* ones, which can ensure a high reflection for *x*-polarized waves. Meanwhile, the reflection-phase $\varphi_{xx}^r(x)$ for this polarization can vary from $-180°$ to $180°$ (Figure 7b) as frequency changes. For the *y*-polarization, however, the coupling between different layers creates a wideband transparent window with a controllable transmission phase, which can again cover the whole 360° range (Figure 7c).

Based on such unique meta-atoms, the authors successfully design several bi-functional meta-devices combining similar or distinct functionalities to manipulate EM waves at different sides of the metasurfaces. For instance, the authors fabricated a bifunctional metasurface, which can anomalously reflect *x*-polarized incident wave to the reflection side (Figure 7d) and focusing the *y*-polarized incident wave to a focal point at the transmission side (Figure 7e). As shown in Figure 7f,g, microwave experiments perfectly demonstrated the desired bi-functionalities of the fabricated meta-device. Both functionalities exhibit very high efficiencies (in the range of 85–91%) and the total thickness of the device is only $\sim\lambda/8$ [97].

Such a strategy has also been extended to design bi-functional PB metasurfaces that can manipulate the wave-fronts of circularly-polarized waves in the full space [89]. The key step is again to find a high-performance PB meta-atom exhibiting unique helicity-dependent transmission/reflection responses. Assisted by an elegant Jones' matrix analysis, Cai et al. were able to find a particular PB meta-atom (see Figure 8a) that can (nearly) perfectly reflect RCP light to RCP light at the reflection side (Figure 8b), and (nearly) transmit LCP light to RCP light at the transmission side (Figure 8c). Such unique polarization-conversion properties make the designed meta-atom ideal candidate to construct bifunctional PB meta-devices for controlling CP lights in the full space [54]. One can simply rotate the meta-atoms with appropriate angles to realize the desired phase distributions to achieve certain functionalities. The authors used such a meta-atom to fabricate a bifunctional meta-lens, and experimentally demonstrated that it behaves as a transmission-mode focusing lens for LCP wave (Figure 8d,f), but changes to a reflection-mode diverging lens for RCP wave (Figure 8e,g). Other bi-functional meta-devices can also be realized. However, one should note that the bi-functional meta-devices realized by such a scheme can only exhibit the same type of wave-manipulation functionalities for two polarizations, since the two phase distributions are intrinsically linked with each other.

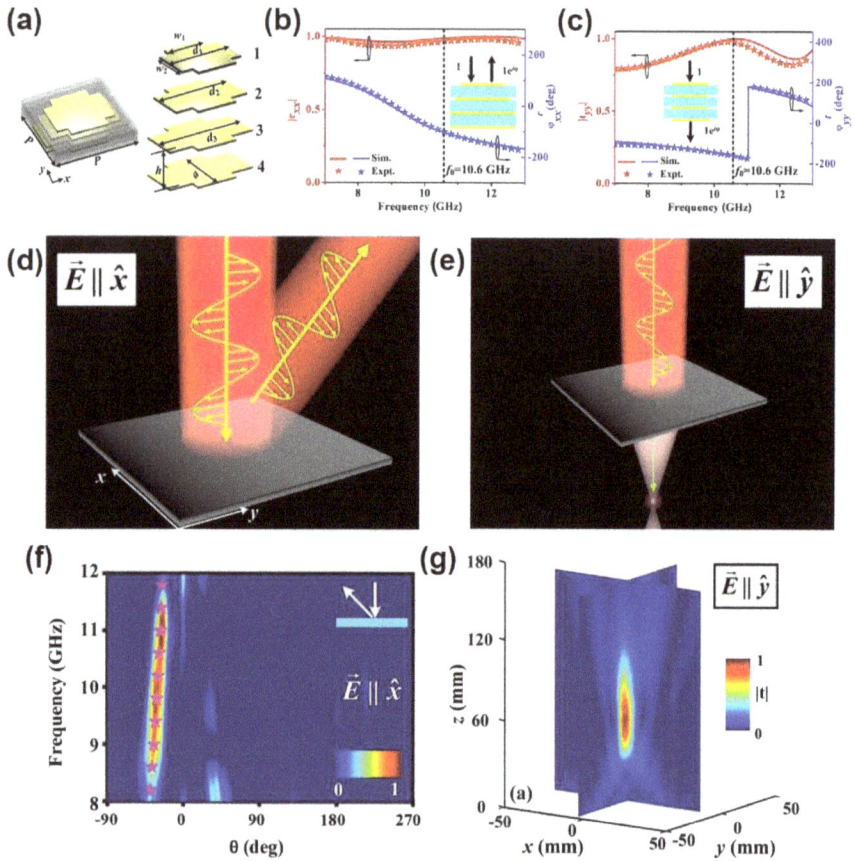

Figure 7. (**a**) Schematics of a 4-layer meta-atom composed by four metallic layers separated by three spacers. Measured and Finite-Difference Time-Domain (FDTD) simulated amplitude-phase spectra of reflection (**b**) and transmission (**c**) for a periodic metasurface made by the meta-atom given in (**a**), under excitations with different polarizations. Schematics of a bi-functional metasurface which behaves as a (**d**) a reflective beam deflector and (**e**) a transmissive lens under excitations of *x*- and *y*-polarized waves, respectively. Measured scattered field intensity versus frequency and detecting angle at reflection sides (**f**) of the metasurface shined by *x*-polarized microwaves and electric field distributions on both *xoz* and *yoz* planes at transmission sides (**g**) of the metasurface shined by *y*-polarized microwaves. Reproduced from [97] with permission.

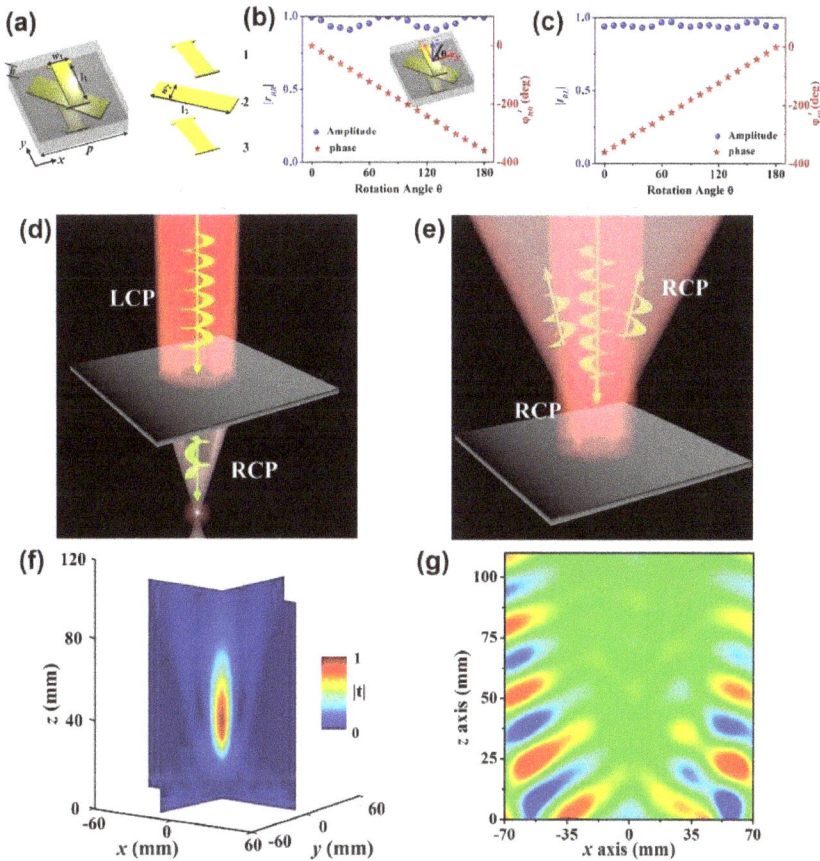

Figure 8. (**a**) Topology of a meta-atom consisting of three metallic layers (yellow) separated by two diecltric spacers (gray). (**b,c**) Amplitude/phase (Blue circle/red star) responses of the meta-atom rotated with certain angles when illuminated by normally incident (**b**) RCP and (**c**) LCP waves. Schematics of a bifunctional metasurface which behaves as a transmissive focusing lens (**d**) and a reflective diverging lens (**e**) under excitations of LCP and RCP waves, respectively. (**f**) Measured $|E_x|^2$ distributions on both *xoz* and *yoz* planes at the transmission side when the meta-device is illuminated by a normally incident LCP wave. (**g**) Measured $\mathrm{Re}(E_x)$ distributions on the *xoz* plane at the reflection side of the metasurface under excitation of a normally-incident RCP wave. Reproduced from [89] with permission.

4. Conclusions and Discussion

In summary, we presented a concise review on the development of multifunctional metasurfaces, focusing on the physical mechanisms and practical meta-devices realizations. The design strategies include a simple "merging" concept and more sophisticated ones relying on single anisotropic meta-atoms, and the realized meta-devices can achieve two or more wave-manipulation functionalities at frequencies ranging from microwave to the visible. In reviewing the development of this field, we find that the meta-atom design is of particular importance, since any new type of high-performance multifunctional meta-atom can surely stimulate a series of meta-devices with diversified functionalities. We already see many interesting new designs in the field (for example, the polarization-dependent full-space meta-atom) which help people realize useful functional devices with more flexibilities in controlling EM waves. However, we also noticed that this field is far from mature, since many new ideas are only realized in the microwave regime, leaving plenty of rooms for researchers working at more challenging frequency domains. We hope that this review can serve as a useful guide to help researchers quickly jump into this field, eventually pushing these conceptual laboratory-level protocol devices to integration-optics platforms.

Before concluding this review, we would like to discuss more on the differences in the design of meta-atoms for microwave, Infrared Radiation (IR), or optical frequencies. Given the different behavior of materials in these frequency ranges, the geometry of the meta-atoms needed to achieve a chosen functionality are necessarily different. For example, whereas in the microwave regime metals behave as perfect electric conductors, the metallic losses can be very significant at IR and optical frequencies. As a result, while meta-atoms with multilayer metallic structures are widely used in designing high-efficiency transmissive multifunctional metasurfaces in low-frequency domains, these meta-atom structures are difficult to be used directly at high frequencies because of the fabrication challenges and the material losses. Instead, simple metallic structures (say, metal bars) are currently used in designing transmission-mode multifunctional meta-devices at high frequencies, although the realized meta-devices suffer from low-efficiency issues. Very recently, all-dielectric metasurfaces began to appear, which seems to be a very promising route to overcome such issue for transmissive multifunctional meta-devices at high frequencies. Along with the fast development in nanofabrication technologies, we expect that more excellent works can appear in this field, generating meta-devices which can eventually be used in practice.

Acknowledgments: This work is supported by the National Natural Science Foundation of China (grant no. 11604167, no. 11474057, no. 11404063, no. 11734007, no. 11674068, and no. 61501499), National Basic Research Program of China (grant no. 2017YFA0303504), K. C. Wong Magna Fund in Ningbo University and the Shanghai Science and Technology Committee (grant no. 16ZR1445200 and no. 16JC1403100).

Author Contributions: Tong Cai, He-Xiu Xu, Qiong He and Shulin Sun contributed related materials; Shiwei Tang and Lei Zhou wrote the paper.

Conflicts of Interest: The authors declare no conflict of interest.

References

1. Shelby, R.A.; Smith, D.R.; Schultz, S. Experimental verification of a negative index of refraction. *Science* **2001**, *292*, 77–79. [CrossRef] [PubMed]
2. Veselago, V.G. The electrodynamics of substances with simultaneously negative values of epsilon and mu. *Soviet Phys. Uspekhi* **1968**, *10*, 509–514. [CrossRef]
3. Pendry, J.B. Negative refraction makes a perfect lens. *Phys. Rev. Lett.* **2000**, *85*, 3966. [CrossRef] [PubMed]
4. Valentine, J.; Zhang, S.; Zentgraf, T.; Ulinavila, E.; Genov, D.A.; Bartal, G.; Zhang, X. Three-dimensional optical metamaterial with a negative refractive index. *Nature* **2008**, *455*, 376. [CrossRef] [PubMed]
5. Liu, Z.; Lee, H.; Xiong, Y.; Sun, C.; Zhang, X. Far-field optical hyperlens magnifying sub-diffraction-limited objects. *Science* **2007**, *315*, 1686. [CrossRef] [PubMed]
6. Fang, N.; Lee, H.; Sun, C.; Zhang, X. Sub-Diffraction-Limited Optical Imaging with a Silver Superlens. *Science* **2005**, *308*, 534–537. [CrossRef] [PubMed]
7. Pendry, J.B.; Schurig, D.; Smith, D.R. Controlling electromagnetic fields. *Science* **2006**, *312*, 1780–1782. [CrossRef] [PubMed]
8. Liu, R.; Ji, C.; Mock, J.J.; Chin, J.Y.; Cui, T.J.; Smith, D.R. Broadband ground-plane cloak. *Science* **2009**, *323*, 366–369. [CrossRef] [PubMed]
9. Ma, H.F.; Cui, T.J. Three-dimensional broadband ground-plane cloak made of metamaterials. *Nat. Commun.* **2010**, *1*, 21. [CrossRef] [PubMed]
10. Sun, W.; He, Q.; Hao, J.; Zhou, L. A transparent metamaterial to manipulate electromagnetic wave polarizations. *Opt. Lett.* **2011**, *36*, 927–929. [CrossRef] [PubMed]
11. Ma, S.; Wang, X.; Luo, W.; Sun, S.; Zhang, Y.; He, Q.; Zhou, L. Ultra-wide band reflective metamaterial wave plates for terahertz waves. *Europhys. Lett.* **2017**, *117*, 37007. [CrossRef]
12. Hao, J.; Ren, Q.; An, Z.; Huang, X.; Chen, Z.; Qiu, M.; Zhou, L. Optical metamaterial for polarization control. *Phys. Rev. A* **2009**, *80*, 23807. [CrossRef]

13. Hao, J.; Yuan, Y.; Ran, L.; Jiang, T.; Kong, J.A.; Chan, C.T.; Zhou, L. Manipulating Electromagnetic Wave Polarizations by Anisotropic Metamaterials. *Phys. Rev. Lett.* **2007**, *99*, 63908. [CrossRef] [PubMed]

14. Hao, J.; Wang, J.; Liu, X.; Padilla, W.J.; Zhou, L.; Qiu, M. High performance optical absorber based on a plasmonic metamaterial. *Appl. Phys. Lett.* **2010**, *96*, 251104. [CrossRef]

15. Song, Z.; He, Q.; Xiao, S.; Zhou, L. Making a continuous metal film transparent via scattering cancellations. *Appl. Phys. Lett.* **2012**, *101*, 5822. [CrossRef]

16. Andryieuski, A.; Lavrinenko, A.V.; Gritti, C.; Zhou, L.; Zalkovskij, M.; Jepsen, P.U.; He, Q.; Malureanu, R.; Song, Z. A new method for obtaining transparent electrodes. *Opt. Express* **2012**, *20*, 22770–22782.

17. Enoch, S.; Tayeb, G.; Sabouroux, P.; Guérin, N.; Vincent, P. A metamaterial for directive emission. *Phys. Rev. Lett.* **2002**, *89*, 213902. [CrossRef] [PubMed]

18. Jiang, Z.H.; Wu, Q.; Werner, D.H. Demonstration of enhanced broadband unidirectional electromagnetic radiation enabled by a subwavelength profile leaky anisotropic zero-index metamaterial coating. *Phys. Rev. B* **2012**, *86*, 125131. [CrossRef]

19. Jeon, S.; Menard, E.; Park, J.; Maria, J.; Meitl, M.; Zaumseil, J.; Rogers, J. Three-dimensional nanofabrication with rubber stamps and conformable photomasks. *Adv. Mater.* **2004**, *16*, 1369–1373. [CrossRef]

20. Rill, M.; Plet, C.; Thiel, M.; Wegener, M.; Freymann, G.V.; Linden, S. Photonic Metamaterials by Direct Laser Writing and Silver Chemical Vapor Deposition. *Nat. Mater.* **2008**, *7*, 543–546. [CrossRef] [PubMed]

21. Ni, X.; Emani, N.K.; Kildishev, A.V.; Boltasseva, A.; Shalaev, V.M. Broadband light bending with plasmonic nanoantennas. *Science* **2012**, *335*, 427. [CrossRef] [PubMed]

22. Pors, A.; Albrektsen, O.; Radko, I.P.; Bozhevolnyi, S.I. Gap plasmon-based metasurfaces for total control of reflected light. *Sci. Rep.* **2013**, *3*, 2155. [CrossRef] [PubMed]

23. Sun, S.; Yang, K.Y.; Wang, C.M.; Juan, T.K.; Chen, W.T.; Liao, C.Y.; He, Q.; Xiao, S.; Kung, W.T.; Guo, G.Y. High-Efficiency Broadband Anomalous Reflection by Gradient Meta-Surfaces. *Nano Lett.* **2012**, *12*, 6223–6229. [CrossRef] [PubMed]

24. Pfeiffer, C.; Grbic, A. Metamaterial Huygens' surfaces: tailoring wave fronts with reflectionless sheets. *Phys. Rev. Lett.* **2013**, *110*, 197401. [CrossRef] [PubMed]

25. Luo, J.; Yu, H.; Song, M.; Zhang, Z. Highly efficient wavefront manipulation in terahertz based on plasmonic gradient metasurfaces. *Opt. Lett.* **2014**, *39*, 2229–2231. [CrossRef] [PubMed]

26. Pfeiffer, C.; Emani, N.K.; Shaltout, A.M.; Boltasseva, A.; Shalaev, V.M.; Grbic, A. Efficient light bending with isotropic metamaterial Huygens' surfaces. *Nano Lett.* **2014**, *14*, 2491. [CrossRef] [PubMed]

27. Wei, Z.; Cao, Y.; Su, X.; Gong, Z.; Long, Y.; Li, H. Highly efficient beam steering with a transparent metasurface. *Opt. Express* **2013**, *21*, 10739. [CrossRef] [PubMed]

28. Yu, N.; Genevet, P.; Kats, M.A.; Aieta, F.; Tetienne, J.P.; Capasso, F.; Gaburro, Z. Light Propagation with Phase Discontinuities Reflection and Refraction. *Science* **2011**, *334*, 333. [CrossRef] [PubMed]

29. Miao, Z.; Wu, Q.; Li, X.; He, Q.; Ding, K.; An, Z.; Zhang, Y.; Zhou, L. Widely Tunable Terahertz Phase Modulation with Gate-Controlled Graphene Metasurfaces. *Phys. Rev. X* **2015**, *5*, 41027. [CrossRef]

30. Sun, S.; He, Q.; Xiao, S.; Xu, Q.; Li, X.; Zhou, L. Gradient-index meta-surfaces as a bridge linking propagating waves and surface waves. *Nat. Mater.* **2012**, *11*, 426–431. [CrossRef] [PubMed]

31. Sun, W.; He, Q.; Sun, S.; Zhou, L. High-efficiency surface plasmon meta-couplers: concept and microwave-regime realizations. *Light Sci. Appl.* **2016**, *5*, e16003. [CrossRef]

32. Lockyear, M.J.; Hibbins, A.P.; Sambles, J.R. Microwave surface-plasmon-like modes on thin metamaterials. *Phys. Rev. Lett.* **2009**, *102*, 73901. [CrossRef] [PubMed]

33. Xiong, X.; Hu, Y.S.; Jiang, S.C.; Hu, Y.H.; Fan, R.H.; Ma, G.B.; Shu, D.J.; Peng, R.W.; Wang, M. Metallic stereostructured layer: An approach for broadband polarization state manipulation. *Appl. Phys. Lett.* **2014**, *105*, 1304. [CrossRef]

34. Ding, F.; Wang, Z.; He, S.; Shalaev, V.M.; Kildishev, A.V. Broadband High-Efficiency Half-Wave Plate: A Super-Cell Based Plasmonic Metasurface Approach. *ACS Nano* **2015**, *9*, 4111–4119. [CrossRef] [PubMed]

35. Yang, Y.; Wang, W.; Moitra, P.; Kravchenko, I.I.; Briggs, D.P.; Valentine, J. Dielectric meta-reflectarray for broadband linear polarization conversion and optical vortex generation. *Nano Lett.* **2014**, *14*, 1394–1399. [CrossRef] [PubMed]

36. Jiang, S. Controlling the Polarization State of Light with a Dispersion-Free Metastructure. *Phys. Rev. X* **2014**, *4*, 21026. [CrossRef]

37. Pfeiffer, C.; Grbic, A. Bianisotropic Metasurfaces for Optimal Polarization Control: Analysis and Synthesis. *Phys. Rev. Appl.* **2014**, *2*, 44011. [CrossRef]
38. Pfeiffer, C.; Zhang, C.; Ray, V.; Guo, L.J.; Grbic, A. High performance bianisotropic metasurfaces: asymmetric transmission of light. *Phys. Rev. Lett.* **2014**, *113*, 23902. [CrossRef] [PubMed]
39. Lee, B.; Yun, H.; Sung, J.; Yun, J.G.; Kim, J.; Lee, K.; Kim, S.J.; Lee, Y. Broadband ultrathin circular polarizer at visible and near-infrared wavelengths using a non-resonant characteristic in helically stacked nano-gratings. *Opt. Express* **2017**, *25*, 14260.
40. Li, X.; Xiao, S.; Cai, B.; He, Q.; Cui, T.J.; Zhou, L. Flat metasurfaces to focus electromagnetic waves in reflection geometry. *Opt. Lett.* **2012**, *37*, 4940. [CrossRef] [PubMed]
41. Aieta, F.; Kats, M.A.; Genevet, P.; Capasso, F. Multiwavelength achromatic metasurfaces by dispersive phase compensation. *Science* **2015**, *347*, 1342–1345. [CrossRef] [PubMed]
42. Ma, X.; Pu, M.; Li, X.; Huang, C.; Wang, Y.; Pan, W.; Zhao, B.; Cui, J.; Wang, C.; Zhao, Z.Y. A planar chiral meta-surface for optical vortex generation and focusing. *Sci. Rep.* **2015**, *5*, 10365. [CrossRef] [PubMed]
43. Huang, L.; Chen, X.; Mühlenbernd, H.; Zhang, H.; Chen, S.; Bai, B.; Tan, Q.; Jin, G.; Cheah, K.; Qiu, C. Three-dimensional optical holography using a plasmonic metasurface. *Nat. Commun.* **2013**, *4*, 2808. [CrossRef]
44. Zheng, G.; Mühlenbernd, H.; Kenney, M.; Li, G.; Zentgraf, T.; Zhang, S. Metasurface holograms reaching 80% efficiency. *Nat. Nanotechnol.* **2015**, *10*, 308. [CrossRef] [PubMed]
45. Lee, G.Y.; Yoon, G.; Lee, S.Y.; Yun, H.; Cho, J.; Lee, K.; Kim, H.; Rho, J.; Lee, B. Complete amplitude and phase control of light using broadband holographic metasurfaces. *Nanoscale* **2018**, *10*, 4237–4245. [CrossRef] [PubMed]
46. Khorasaninejad, M.; Chen, W.T.; Devlin, R.C.; Oh, J.; Zhu, A.Y.; Capasso, F. Metalenses at visible wavelengths: Diffraction-limited focusing and subwavelength resolution imaging. *Science* **2016**, *352*, 1190. [CrossRef] [PubMed]
47. Arbabi, A.; Horie, Y.; Ball, A.J.; Bagheri, M.; Faraon, A. Subwavelength-thick lenses with high numerical apertures and large efficiency based on high-contrast transmitarrays. *Nat. Commun.* **2014**, *6*, 7069. [CrossRef] [PubMed]
48. Chen, X.; Chen, M.; Mehmood, M.Q.; Wen, D.; Yue, F.; Qiu, C.W.; Zhang, S. Longitudinal Multifoci Metalens for Circularly Polarized Light. *Adv. Opt. Mater.* **2015**, *3*, 1201–1206. [CrossRef]
49. Pfeiffer, C.; Grbic, A. Cascaded metasurfaces for complete phase and polarization control. *Appl. Phys. Lett.* **2013**, *102*, 231116. [CrossRef]
50. Ho, J.S.; Qiu, B.; Tanabe, Y.; Yeh, A.J.; Fan, S.; Poon, A.S.Y. Planar immersion lens with metasurfaces. *Phys. Rev. B* **2015**, *91*, 125145. [CrossRef]
51. Ju, Y.K.; Kim, H.; Kim, B.H.; Chang, T.; Lim, J.; Jin, H.M.; Mun, J.H.; Choi, Y.J.; Chung, K.; Shin, J. Highly tunable refractive index visible-light metasurface from block copolymer self-assembly. *Nat. Commun.* **2016**, *7*, 12911.
52. Park, J.; Kang, J.; Kim, S.J.; Liu, X.; Brongersma, M.L. Dynamic Reflection Phase and Polarization Control in Metasurfaces. *Nano Lett.* **2016**, *17*, 407. [CrossRef] [PubMed]
53. Shi, Y.; Fan, S. Dynamic non-reciprocal meta-surfaces with arbitrary phase reconfigurability based on photonic transition in meta-atoms. *Appl. Phys. Lett.* **2016**, *108*, 1232009–1232150. [CrossRef]
54. Luo, W.; Xiao, S.; He, Q.; Sun, S.; Zhou, L. Photonic Spin Hall Effect with Nearly 100% Efficiency. *Adv. Opt. Mater.* **2015**, *3*, 1102–1108. [CrossRef]
55. Yin, X.; Ye, Z.; Rho, J.; Wang, Y.; Zhang, X. Photonic spin Hall effect at metasurfaces. *Science* **2013**, *339*, 1405–1407. [CrossRef] [PubMed]
56. Kildishev, A.; Shaltout, A.; Liu, J.; Shalaev, V. Photonic spin Hall effect in gap–plasmon metasurfaces for on-chip chiroptical spectroscopy. *Optica* **2015**, *2*, 860.
57. Yue, F.; Wen, D.; Zhang, C.; Gerardot, B.D.; Wang, W.; Zhang, S.; Chen, X. Multichannel Polarization-Controllable Superpositions of Orbital Angular Momentum States. *Adv. Mater.* **2017**, *29*, 1603838. [CrossRef] [PubMed]
58. Mehmood, M.Q.; Mei, S.; Hussain, S.; Huang, K.; Siew, S.Y.; Zhang, L.; Zhang, T.; Ling, X.; Liu, H.; Teng, J. Visible-Frequency Metasurface for Structuring and Spatially Multiplexing Optical Vortices. *Adv. Mater.* **2016**, *28*, 2533–2539. [CrossRef] [PubMed]

59. Chen, W.T.; Yang, K.Y.; Wang, C.M.; Huang, Y.W.; Sun, G.; Chiang, I.; Liao, C.Y.; Hsu, W.L.; Lin, H.T.; Sun, S. High-Efficiency Broadband Meta-Hologram with Polarization-Controlled Dual Images. *Nano Lett.* **2014**, *14*, 225–230. [CrossRef] [PubMed]

60. Wan, X.; Shen, X.; Luo, Y.; Cui, T.J. Planar bifunctional Luneburg-fisheye lens made of an anisotropic metasurface. *Laser Photonics Rev.* **2014**, *8*, 757–765. [CrossRef]

61. Huang, Y.W.; Wei, T.C.; Tsai, W.Y.; Wu, P.C.; Wang, C.M.; Sun, G.; Tsai, D.P. Aluminum Plasmonic Multicolor Meta-Hologram. *Nano Lett.* **2015**, *15*, 3122. [CrossRef] [PubMed]

62. Wen, D.; Chen, S.; Yue, F.; Chan, K.; Chen, M.; Ardron, M.; Li, K.F.; Wong, P.W.H.; Cheah, K.W.; Pun, E.Y.B. Metasurface Device with Helicity-Dependent Functionality. *Adv. Opt. Mater.* **2016**, *4*, 321–327. [CrossRef]

63. Wen, D.; Yue, F.; Li, G.; Zheng, G.; Chan, K.; Chen, S.; Chen, M.; Li, K.F.; Wong, P.W.H.; Cheah, K.W. Helicity multiplexed broadband metasurface holograms. *Nat. Commun.* **2015**, *6*, 8241. [CrossRef] [PubMed]

64. Huang, L.; Mühlenbernd, H.; Li, X.; Song, X.; Bai, B.; Wang, Y.; Zentgraf, T. Broadband hybrid holographic multiplexing with geometric metasurfaces. *Adv. Mater.* **2015**, *27*, 6444–6449. [CrossRef] [PubMed]

65. Wen, D.; Yue, F.; Ardron, M.; Chen, X. Multifunctional metasurface lens for imaging and Fourier transform. *Sci. Rep.* **2016**, *6*, 27628. [CrossRef] [PubMed]

66. Liu, S.; Zhang, L.; Yang, Q.L.; Xu, Q.; Yang, Y.; Noor, A.; Zhang, Q.; Iqbal, S.; Wan, X.; Tian, Z.; et al. Frequency-Dependent Dual-Functional Coding Metasurfaces at Terahertz Frequencies. *Adv. Opt. Mater.* **2016**, *4*, 1965–1973. [CrossRef]

67. Wang, X.; Ding, J.; Zheng, B.; An, S.; Zhai, G.; Zhang, H. Simultaneous Realization of Anomalous Reflection and Transmission at Two Frequencies using Bi-functional Metasurfaces. *Sci. Rep.* **2018**, *8*, 1876. [CrossRef] [PubMed]

68. Stewart, J.W.; Akselrod, G.M.; Smith, D.R.; Mikkelsen, M.H. Toward Multispectral Imaging with Colloidal Metasurface Pixels. *Adv. Mater.* **2017**, *29*, 1602971. [CrossRef] [PubMed]

69. Veksler, D.; Maguid, E.; Shitrit, N.; Ozeri, D.; Kleiner, V.; Hasman, E. Multiple Wavefront Shaping by Metasurface Based on Mixed Random Antenna Groups. *ACS Photonics* **2015**, *2*, 661–667. [CrossRef]

70. Xiang, N.; Cheng, Q.; Chen, H.B.; Zhao, J.; Jiang, W.X.; Ma, H.F.; Cui, T.J. Bifunctional metasurface for electromagnetic cloaking and illusion. *Appl. Phys. Express* **2015**, *8*, 92601. [CrossRef]

71. Chen, X.; Huang, L.; Mühlenbernd, H.; Li, G.; Bai, B.; Tan, Q.; Jin, G.; Qiu, C.; Zhang, S.; Zentgraf, T. Dual-polarity plasmonic metalens for visible light. *Nat. Commun.* **2012**, *3*, 1198. [CrossRef] [PubMed]

72. Pors, A.; Nielsen, M.G.; Bernardin, T.; Weeber, J.; Bozhevolnyi, S.I. Efficient unidirectional polarization-controlled excitation of surface plasmon polaritons. *Light Sci. Appl.* **2014**, *3*, e197. [CrossRef]

73. Hou, H.; Wang, G.; Li, H.; Guo, W.; Li, T. Highly efficient multifunctional metasurface for high-gain lens antenna application. *Appl. Phys. A* **2017**, *123*, 460. [CrossRef]

74. Zhang, C.; Yue, F.; Wen, D.; Chen, M.; Zhang, Z.; Wang, W.; Chen, X. Multichannel metasurface for simultaneous control of holograms and twisted light beams. *ACS Photonics* **2017**, *4*, 1906–1912. [CrossRef]

75. Li, Y.; Li, X.; Chen, L.; Pu, M.; Jin, J.; Hong, M.; Luo, X. Orbital angular momentum multiplexing and demultiplexing by a single metasurface. *Adv. Opt. Mater.* **2017**, *5*, 1600502. [CrossRef]

76. Chen, X.; Huang, L.; Mühlenbernd, H.; Li, G.; Bai, B.; Tan, Q.; Jin, G.; Qiu, C.W.; Zentgraf, T.; Zhang, S. Reversible Three-Dimensional Focusing of Visible Light with Ultrathin Plasmonic Flat Lens. *Adv. Opt. Mater.* **2013**, *1*, 517–521. [CrossRef]

77. Wang, Z.; Jia, H.; Yao, K.; Cai, W.; Chen, H.; Liu, Y. Circular dichroism metamirrors with near-perfect extinction. *ACS Photonics* **2016**, *3*, 2096–2101. [CrossRef]

78. Cai, T.; Wang, G.M.; Zhang, X.F.; Liang, J.G.; Zhuang, Y.Q.; Liu, D.; Xu, H.X. Ultra-Thin Polarization Beam Splitter Using 2-D Transmissive Phase Gradient Metasurface. *IEEE Trans. Antennas Propag.* **2015**, *63*, 5629–5636. [CrossRef]

79. Ma, H.F.; Wang, G.Z.; Kong, G.S.; Cui, T.J. Independent controls of differently-polarized reflected waves by anisotropic metasurfaces. *Sci. Rep.* **2015**, *5*, 9605. [CrossRef] [PubMed]

80. Liu, S.; Cui, T.J.; Xu, Q.; Bao, D.; Du, L.; Wan, X.; Tang, W.X.; Ouyang, C.; Zhou, X.Y.; Yuan, H. Anisotropic coding metamaterials and their powerful manipulation of differently polarized terahertz waves. *Light Sci. Appl.* **2016**, *5*, e16076. [CrossRef]

81. Jia, S.L.; Wan, X.; Bao, D.; Zhao, Y.J.; Cui, T.J. Independent controls of orthogonally polarized transmitted waves using a Huygens metasurface. *Laser Photonics Rev.* **2015**, *9*, 545–553. [CrossRef]

82. Zhu, A.Y.; Kuznetsov, A.I.; Luk Yanchuk, B.; Engheta, N.; Genevet, P. Traditional and emerging materials for optical metasurfaces. *Nanophotonics* **2016**, *6*, 452–471. [CrossRef]
83. Mosallaei, H.; Farmahinifarahani, M. Birefringent reflectarray metasurface for beam engineering in infrared. *Opt. Lett.* **2013**, *38*, 462.
84. Boroviks, S.; Deshpande, R.A.; Mortensen, N.A.; Bozhevolnyi, S.I. Multifunctional meta-mirror: polarization splitting and focusing. *ACS Photonics* **2017**. [CrossRef]
85. Maguid, E.; Yulevich, I.; Veksler, D.; Kleiner, V.; Brongersma, M.L.; Hasman, E. Photonic spin-controlled multifunctional shared-aperture antenna array. *Science* **2016**, *352*, 1202. [CrossRef] [PubMed]
86. Lin, D.; Holsteen, A.L.; Maguid, E.; Wetzstein, G.; Kik, P.G.; Hasman, E.; Brongersma, M.L. Photonic Multitasking Interleaved Si Nanoantenna Phased Array. *Nano Lett.* **2016**, *16*, 7671. [CrossRef] [PubMed]
87. Song, E.Y.; Lee, G.Y.; Park, H.; Lee, K.; Kim, J.; Hong, J.; Kim, H.; Lee, B. Compact Generation of Airy Beams with C-Aperture Metasurface. *Adv. Opt. Mater.* **2017**, *5*, 1601028. [CrossRef]
88. Ling, X.; Zhou, X.; Shu, W.; Luo, H.; Wen, S. Realization of Tunable Photonic Spin Hall Effect by Tailoring the Pancharatnam-Berry Phase. *Sci. Rep.* **2014**, *4*, 5557. [CrossRef] [PubMed]
89. Cai, T.; Wang, G.; Xu, H.; Tang, S.; Li, H.; Liang, J.; Zhuang, Y. Bifunctional Pancharatnam-Berry metasurface with high-efficiency helicity-dependent transmissions and reflections. *Ann. Phys.* **2018**, *530*, 1700321. [CrossRef]
90. Cai, T.; Wang, G.; Xu, H.; Tang, S.; Liang, J. Polarization-independent broadband meta-surface for bifunctional antenna. *Opt. Express* **2016**, *24*, 22606–22615. [CrossRef] [PubMed]
91. Ma, S.; Xiao, S.; Zhou, L. Resonant modes in metal/insulator/metal metamaterials: An analytical study on near-field couplings. *Phys. Rev. B* **2016**, *93*, 45305. [CrossRef]
92. Qu, C.; Ma, S.; Hao, J.; Qiu, M.; Li, X.; Xiao, S.; Miao, Z.; Dai, N.; He, Q.; Sun, S. Tailor the Functionalities of Metasurfaces Based on a Complete Phase Diagram. *Phys. Rev. Lett.* **2015**, *115*, 235503. [CrossRef] [PubMed]
93. Cai, T.; Tang, S.; Wang, G.; Xu, H.; Sun, S.; He, Q.; Zhou, L. High-performance bifunctional metasurfaces in transmission and reflection geometries. *Adv. Opt. Mater.* **2017**, *5*, 1600506. [CrossRef]
94. Ding, F.; Deshpande, R.; Bozhevolnyi, S.I. Bifunctional Gap-Plasmon Metasurfaces for Visible Light: Polarization-Controlled Unidirectional Surface Plasmon Excitation and Beam Steering at Normal Incidence. *Light Sci. Appl.* **2018**. [CrossRef]
95. Xu, H.X.; Tang, S.; Ling, X.; Luo, W.; Zhou, L. Flexible control of highly-directive emissions based on bifunctional metasurfaces with low polarization cross-talking. *Ann. Phys.* **2017**, *529*, 1700045. [CrossRef]
96. Chen, K.; Feng, Y.; Monticone, F.; Zhao, J.; Zhu, B.; Jiang, T.; Zhang, L.; Kim, Y.; Ding, X.; Zhang, S. A Reconfigurable Active Huygens' Metalens. *Adv. Mater.* **2017**, *29*, 1606422. [CrossRef] [PubMed]
97. Cai, T.; Wang, G.; Tang, S.; Xu, H.; Duan, J.; Guo, H.; Guan, F.; Sun, S.; He, Q.; Zhou, L. High-efficiency and full-space manipulation of electromagnetic wave fronts with metasurfaces. *Phys. Rev. Appl.* **2017**, *8*, 34033. [CrossRef]

applied sciences

MDPI

Review

Orbital Angular Momentum Generation and Detection by Geometric-Phase Based Metasurfaces

Menglin L. N. Chen [1,2], Li Jun Jiang [1,2,*] and Wei E. I. Sha [3,*]

[1] Department of Electrical and Electronic Engineering, The University of Hong Kong, Hong Kong, China; menglin@connect.hku.hk
[2] HKU Shenzhen Institute of Research and Innovation, Shenzhen 518057, China
[3] Key Laboratory of Micro-Nano Electronic Devices and Smart Systems of Zhejiang Province, College of Information Science and Electronic Engineering, Zhejiang University, Hangzhou 310027, China
* Correspondence: jianglj@hku.hk (L.J.J.); weisha@zju.edu.cn (W.E.I.S.)

Received: 26 January 2018; Accepted: 18 February 2018; Published: 2 March 2018

Abstract: We present a comprehensive review on the geometric-phase based metasurfaces for orbital angular momentum (OAM) generation and detection. These metasurfaces manipulate the electromagnetic (EM) wave by introducing abrupt phase change, which is strongly dependent on the polarization state of incident EM wave and can be interpreted by geometric phase. Hence, the conventional bulk devices that based on the accumulated phase change along the optical path can be avoided.

Keywords: metasurfaces; geometric phase; orbital angular momentum (OAM)

1. Introduction

It is well known that electromagnetic (EM) wave carries linear momentum which is associated with the poynting vector. Besides the linear momentum, EM wave has been demonstrated to possess angular momentum (AM), including spin angular momentum (SAM) and orbital angular momentum (OAM) [1]. Circularly polarized wave carries SAM. The SAM is $\hbar$ per photon for left-handed circular polarization (LHCP) and $-\hbar$ per photon for right-handed circular polarization (RHCP), where $\hbar$ is the reduced Planck constant. The SAM characterizes the spin feature of photon. Different from the SAM, OAM manifests the orbital rotation of photon and each photon has an OAM of $l\hbar$, where l is known as the OAM index and can be any integer. Different values of l correspond to mutually orthogonal OAM states and the number of allowable OAM states for photons is unbounded. Different OAM states have been used to encode information in communications to enhance the channel capacity, both in free space [2–9] and optical fibers [10]. The main problem of practical application of OAM in communications is the significant crosstalk between OAM modes [11,12]. The spatial-dependence and divergency nature of OAM-carrying waves result in the their vulnerability to the atmosphere [13,14] and limited power received at the receiver side [15]. All these facts will degrade the purity of OAM modes. Still, OAM offers a new attractive degree of freedom to EM waves, extending beyond the existing wave features. The optical vortices of OAM beams also find their applications in super resolution imaging [16,17]. OAM can be transferred to particles, which has been applied in optical tweezers [18–20]. The exchange of OAM with matter can also be utilized in detecting the rotation of particles [21]. In quantum mechanics, at a single-photon level, OAM modes can be utilized for high-dimensional entanglement [22,23]. Overall, OAM has received great attentions in multidisciplinary research areas [24–27].

Since OAM has great potentials in various applications, research has been undertaken extensively on its generation. In 1992, Allen et al. first found that a Laguerre-Gaussian (LG) beam with helical wavefront carries well defined OAM [28]. That beam possesses an azimuthal phase term, $e^{il\phi}$, where ϕ is the azimuthal angle. Before that, Soskin et al. created structured light with helical wavefront by forked gratings [29]. This grating forms the basis of the computer generated holograms (CGHs) for producing light beams with OAM. Afterwards, many other devices for OAM generation have been proposed, such as the cylindrical lens in 1993 [30], spiral phase plates (SPPs) in 1994 [31], q plates in 2006 [32]. Lately, OAM has been analyzed at radio frequencies. Antenna arrays [33], traveling-wave antennas [34] and circularly polarized patch antennas [35] which radiate EM waves carrying OAM have been demonstrated.

Recent advances in versatile metasurfaces also expedite powerful and convenient design routes for OAM generation [36–39]. The concept of metasurface originates from the conventional frequency selective surfaces (FSSs) [40]. They are composed of man-made subwavelength scatterers with varying geometry and orientation. Therefore, they go beyond the conventional FSSs due to the high feasibility of tailoring the geometry and orientation of the ultracompact scatterers. Metasurfaces locally alter the wave properties by the abrupt phase change at the scatterers. By varying the geometry or orientation, scatterers can cover a total 2π phase shift so that arbitrary beam forming can be achieved. Meanwhile, scatterers can be designed to simultaneously change the wave amplitude [41]. Besides the electric response, they can have the magnetic response. Scatterers have both electric and magnetic responses form the Huygens metasurface [42]. The magnetic response helps to compensate the impedance mismatch at the metasurface interface so that nearly perfect efficiency can be achieved.

Although there are various designs in different types and suitable for different application scenarios, the design principles for OAM generation lead to one common rule: the introduction of the azimuthal phase term $e^{il\phi}$ to EM waves. Generally, they fall into two categories: independent and dependent on the wave polarization.

The first scheme employs isotropic materials, such as SPPs, CGHs. $e^{il\phi}$ is introduced in a SPP based on the accumulated spatially varying optical path [43–45]. They are usually implemented at optical frequencies. At lower frequencies, the device would become bulky unless some flatten techniques are employed [46–48]. CGHs are diffractive optical elements and they produce light with different OAMs according to the diffraction orders [49–52]. Alternatively, the excitation of an antenna or antenna array can be modulated directly to satisfy the required phase condition so that an OAM wave can be radiated out [53,54]. $e^{il\phi}$ can also be produced based on the abrupt phase shift at scatterers on a metasurface. By varying the geometry of the scatterer, its resonant frequency is changed so that the phase shift varies at the designed frequency. A total 2π phase shift is achieved after optimization; and successful generation of different OAM states has been reported in [55,56]. Lately, tunable scatterers loaded by varactor diodes are proposed for convenient multiple OAM-mode generation [57]. Scatterers can also be made anisotropic to achieve independent control of different polarizations. However, even if the response of scatterers is polarization dependent, the helicity of the produced OAM does not depend on the polarization state of incident wave. In other words, the helicity is fixed. While for the second scheme, it is an opposite scenario.

The second scheme is based on the conversion and coupling between SAM and OAM. This process occurs in inhomogeneous and anisotropic media, such as q plates [58,59]. Q plates shift the circular polarization state from left to right or right to left and have spatially varying optical axis [60]. Their behaviors can be explained by AM conservation law. The flip of the circular polarization state indicates a change of $\pm 2\hbar$ in SAM. When the q plate is cylindrically symmetric, the total AM is conserved so that the output wave must carry an OAM of $\mp 2\hbar$. When the q plate is not cylindrically symmetric, it introduces extra AM to the system so that different orders of OAM can be generated in the output wave. Q plates are usually made by liquid crystals, owning to their flexibility and anisotropic properties. By tuning the temperature or external voltage, the birefringent retardation in liquid crystals varies, so that the conversion efficiency can be tuned [61,62]. The maximum efficiency is obtained when the retardation is π. Metasurfaces have been used to implement the feature of q plates [63–67]. The flip of the circular polarization state achieved by anisotropic scatterers and rotation of the scatterers is an analogue to the rotation of optical axis of a q plate, along with the introduction of geometric phase. Therefore, this type of metasurfaces is known as geometric-phase metasurfaces. Unlike the metasurfaces employing the first scheme, geometric-phase metasurfaces produce OAM along with the change of the polarization state of the output wave, which is fundamentally different from the first scheme. The handness of the produced OAM depends on the incident SAM. Hence, geometric-phase metasurfaces show their high flexibility in the manipulation of OAM because they enable the coupling and interchange between SAM and OAM.

In communications, detection of OAM at the receiver side is required. Although OAM detection is just a reciprocal process of OAM generation, it is much more challenging due to the degraded OAM states after the propagation. OAM detection can be classified into three categories, mode analysis based on field data [68–70], observation of OAM induced effects such as the rotational Doppler shift [71–73] and the beam reforming by adopting holographic technology [74]. The first approach needs to acquire all the three components of electric and magnetic fields so that the OAM can be calculated explicitly [68]. Alternatively, one can measure the phase gradient that is equal to the OAM index l [75]. In the third approach, by using a hologram, the OAM modes can be projected into a detectable Gaussian mode [76]. Geometric-phase metasurfaces can be used to implement the holograms and have the advantages of low profile and high tuning flexibility.

In this review article, we concentrate on the current research of the OAM generation and detection by geometric-phase based metasurfaces. We start by the introduction of geometric phase. Then, we go over several novel metasurfaces for OAM generation. The induced geometric phases on the metasurfaces can be in both discrete and continuous formats. Their geometries, working principles and novel functionalities have been reviewed and discussed in detail. At last, a method for OAM detection is reviewed.

2. Geometric Phase

When a light changes its initial polarization state to a final polarization state along different paths on a Poincaré sphere, the final polarization states will have difference phases, known as geometric phase [77]. In the following, we derive the geometric phase when the SAM shifts from $\pm\hbar$ to $\mp\hbar$ through an anisotropic scatterer with different orientations.

The behavior of a scatterer can be modelled by Jones matrix **J**. It connects the polarization state of the scattered wave with that of an incident wave:

$$\begin{pmatrix} E_x^s \\ E_y^s \end{pmatrix} = \begin{pmatrix} J_{xx} & J_{xy} \\ J_{yx} & J_{yy} \end{pmatrix} \begin{pmatrix} E_x^i \\ E_y^i \end{pmatrix} = \mathbf{J}^l \begin{pmatrix} E_x^i \\ E_y^i \end{pmatrix}, \tag{1}$$

where E_x^i and E_y^i are the x and y components of the incident electric field. E_x^s and E_y^s are the corresponding components of the scattered electric field.

Jones matrix under the circular basis can be obtained by coordinate transformation,

$$\mathbf{J}^c = \begin{pmatrix} J_{++} & J_{+-} \\ J_{-+} & J_{--} \end{pmatrix} = \frac{1}{2} \begin{pmatrix} (J_{xx} + J_{yy}) + i(J_{xy} - J_{yx}) & (J_{xx} - J_{yy}) - i(J_{xy} + J_{yx}) \\ (J_{xx} - J_{yy}) + i(J_{xy} + J_{yx}) & (J_{xx} + J_{yy}) - i(J_{xy} - J_{yx}) \end{pmatrix}, \tag{2}$$

where $+$ and $-$ represent the right circularly polarized (RCP) and left circularly polarized (LCP) components.

If the scatterer achieves a perfect conversion between RCP wave and LCP wave, it must satisfy $J_{yy} = -J_{xx}$ and $J_{yx} = J_{xy} = 0$, so that the Jones matrix is written as,

$$\mathbf{J}_{pol}^l = \begin{pmatrix} J_{xx} & 0 \\ 0 & -J_{xx} \end{pmatrix}, \quad \mathbf{J}_{pol}^c = \begin{pmatrix} 0 & J_{xx} \\ J_{xx} & 0 \end{pmatrix} \tag{3}$$

Clearly, there are only off-diagonal entities in $\mathbf{J}_{pol}^c$, indicating the flip between RHCP and LHCP. Axially rotating the scatterer by an angle of α will result in a new Jones matrix,

$$\mathbf{J}_{pol}^l(\alpha) = J_{xx} \begin{pmatrix} \cos(2\alpha) & \sin(2\alpha) \\ \sin(2\alpha) & -\cos(2\alpha) \end{pmatrix}, \quad \mathbf{J}_{pol}^c(\alpha) = J_{xx} \begin{pmatrix} 0 & e^{-2i\alpha} \\ e^{2i\alpha} & 0 \end{pmatrix} \tag{4}$$

$\mathbf{J}_{pol}^c(\alpha)$ also has only the non-zero off-diagonal items. Besides, an additional phase factor $e^{\pm 2i\alpha}$ is introduced. This phase is the so-called geometric phase. The scatterers on a geometric-phase metasurface cover a 2π phase shift by changing the rotation angle α.

To generate OAM, geometric phase is utilized to construct the required phase profile, $e^{il\phi}$ of an OAM wave. Apparently, to generate an OAM of order l, scatterers on a metasurface with azimuthal location ϕ should be designed to provide a phase change of $l\phi$, i.e., $\alpha = \pm l\phi/2$. The sign depends on the incident circular polarization state. The value of α/ϕ is known as the topology charge q of the metasurface. Apparently, the OAM order l is determined by q and should be double of its value.

The geometric phase itself does not depend on the frequency so that broadband metasurfaces could be potentially designed. For a perfect (100%) conversion, the condition $J_{yy} = -J_{xx} = \pm 1$ and $J_{yx} = J_{xy} = 0$ should be strictly satisfied, i.e., J_{yy} and J_{xx} should have the same unit amplitude and π phase difference. Achieving these conditions by scatterers on resonance will inevitably limit the bandwidth. When the requirement is not strictly satisfied, the co-circularly polarized component needs to be filtered out. Only the cross-circularly polarized component carries OAM.

3. Metasurfaces for OAM Generation

Various prototypes of scatterers that can totally or partially convert the R/LHCP to L/RHCP have been proposed for OAM generation. Plasmonic scatterers, such as the golden nanorods in [78], L-shaped gold nanoantennas [79], rectangular split-ring resonators (SRRs) in [80]; apertures opened on metals, such as the elliptical nanoholes in [81]; and dielectric particles have been demonstrated at optical regime. At microwave frequencies, multi-layered unit cells have been reported to achieve high efficiency [82,83]. The scatterers and their composite metasurfaces are usually simulated by numerical simulation software, such as finite element method and finite integration technique-based CST Microwave Studio [84–89], finite element method-based HFSS, finite-difference time-domain method-based Lumerical [90], and COMSEL. Some of them can be analytically modelled [87] or simulated by approximate models [84,89]. Lately, an efficient modelling is applied to integral equation solvers to accelerate the simulation of metasurfaces in multiscale [91].

3.1. Artifical PEC-PMC Metasurface

We propose a composite perfect electric conductor (PEC)-perfect magnetic conductor (PMC)-based anisotropic metasurface [84]. A PEC surface is designed for x polarization so we have $J_{xx} = -1$. An artificial PMC surface is put beneath the PEC surface so that $J_{yy} = 1$ is satisfied. The proposed novel metasurface is depicted in Figure 1. The metasurface has two dielectric layers and a ground plane. It is composed of scatterers shown in Figure 1b. The metal strips on the top layer function as a parallel plate waveguide and allow the perfect transmission of y polarized wave while the x polarized wave will be totally reflected with π phase shift. The metal patch in middle together with the via and ground plane forms the well-known mushroom-like high-impedance surface. By operating at its resonant frequency, it can be considered as a PMC surface and will reflect the transmitted y polarized wave from the metal strips with a zero phase shift. The whole metasurface is built by distributing the scatterers with varying orientations according to their azimuthal locations. It should be noted that the mushroom-like high-impedance surface is isotropic, therefore, there is no need to change the orientations of mushrooms. For the PEC surface, by setting the rotation angle $\alpha = \phi$, rotation of the strips will lead to a series of concentric loops. The resultant generated OAM order is 2 or -2, where the sign depends on the incident circular polarization state.

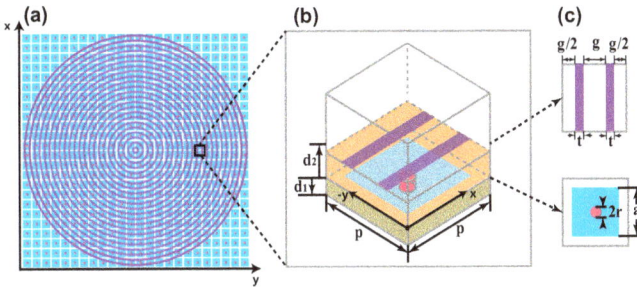

Figure 1. Schematic pattern of the perfect electric conductor (PEC)-perfect magnetic conductor (PMC) anisotropic metasurface for orbital angular momentum (OAM) generation. With a nearly 100% conversion efficiency, the metasurface perfectly converts a left (right) circularly polarized plane wave carrying zero OAM to a right (left) circularly polarized vortex beam carrying $\pm 2\hbar$ OAM: (**a**) Top view of the whole metasurface; (**b,c**) A scatterer in the metasurface. The scatterer is composed of artificial PEC (purple) and PMC (blue and red) surfaces. The period of the scatterer is $p = 7$ mm. The permittivity and thickness of the dielectric substrate are set to $\epsilon_r = 2.2$, $d_1 = 2$ mm and $d_2 = 3$ mm. For the artificial PEC surface (top-right inset), the width and gap for the strip is $t = 1$ mm and $g = 2.5$ mm, respectively. For the mushroom-based artificial PMC surface (bottom-right inset), the square patch size is $a = 6$ mm. A metallic via with the radius of $r = 0.25$ mm and height of $d_1 = 2$ mm connects the patch to the ground plane. Reproduced with permission from [84], Copyright AIP Publishing LLC, 2016.

The metasurface is designed at 6.2 GHz and can be conveniently fabricated using printed circuit board (PCB) technique. Full-wave simulation was done in CST MWS. The right circularly polarized plane wave is used as incident wave. Figure 2 shows the amplitude and phase distributions of the reflected electric fields at a transverse plane of $z = 20$ mm. An amplitude null can be observed in Figure 2a, which results from the phase singularity at the center for an OAM-carrying wave. In Figure 2b, the phase accumulated along a full circular path around the center is 4π, indicating an OAM of order -2. A unique feature for the PEC-PMC metasurface is that the scatterers on top layer are not on resonance or discrete but continuously connected. Thus, the near-field pattern is quite smooth without any evanescent field component scattered by discrete scatterers.

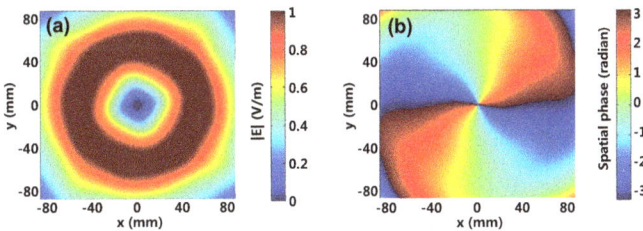

Figure 2. The amplitude and phase distributions of reflected electric fields from the PEC-PMC metasurface at a transverse plane $z = 20$ mm: (**a**) Amplitude; (**b**) Phase. Reproduced with permission from [84], Copyright AIP Publishing LLC, 2016.

For the generation of OAM of other orders, a similar PEC-PMC metasurface but with discrete dipole scatterers on the top layer is proposed. The dipoles are designed to totally reflect the x polarized wave with π phase shift at 6.2 GHz. Two cases for generation of OAM of order -2 and -4 are shown in Figure 3. The amplitude and phase of the reflected field are drawn at a transverse plane of $z = 40$ mm. The distributed dipole scatterers are also depicted in Figure 3a,c. The field patterns verify the desired OAMs have been generated. However, due to the influence of the discrete dipoles, the amplitude distribution is not uniform any more and ripples are observed in the phase distribution. As discussed before, the ripples are caused by evanescent field components scattered by the dipoles. The distortion only exists in the near field and will become less serious when the observation plane moves further away from the metasurface.

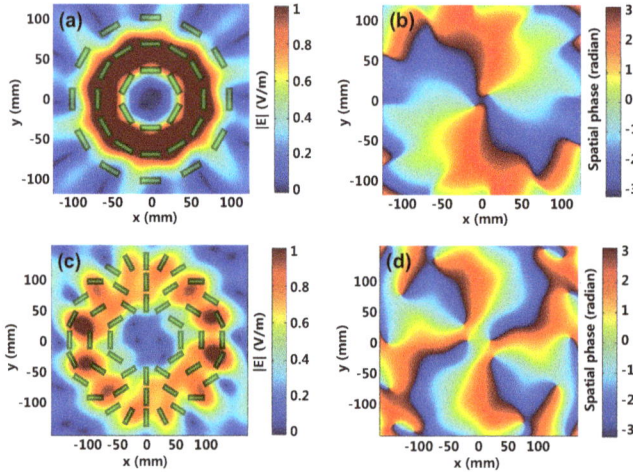

Figure 3. The amplitude and phase distributions of reflected electric fields from the discrete PEC-PMC metasurface. For the generation of OAM of order -2, (**a**) amplitude; (**b**) phase at a transverse plane $z = 40$ mm. For the generation of OAM of order -4, (**c**) amplitude; (**d**) phase at a transverse plane $z = 100$ mm. Reproduced with permission from [84], Copyright AIP Publishing LLC, 2016.

3.2. Ultrathin Complementary Metasurface

The efficiency of reflective metasurface could be very high by placing a reflector beneath it, while for transmissive metasurface, high efficiency cannot be easily achieved due to the reflection at the metasurface-air interface stemming from the impedance mismatch. In the following, we show our designed double-layer complementary metasurface for OAM generation with high efficiency [85].

The unit cell is shown in Figure 4a. Each unit cell consists of four complementary split-ring resonators (CSRRs) with two different sizes and orientations. The equivalent circuit and its response for each pair are drawn in Figure 4b. It is clear that the double-layer structure offers two parallel inductor-capacitor (LC) resonant circuits so that we have a sufficient tunable range to achieve the design objective, i.e., the same high transmittance from the two pairs as well as the π phase difference between their transmission coefficients. Figure 4c presents the full-wave simulation results of unit cell. At 17.85 GHz, magnitudes of the co-transmission coefficients are both 0.91 and their phase difference is π, indicating a 81% circular polarization conversion efficiency. This double-layer structure can be fabricated on a conventional PCB.

Then a whole metasurface is built by arranging the unit cells with varying orientations. Two metasurfaces with topological charges of 1 and 2 are designed. The top view of the designed metasurface is shown in Figure 5. Each metasurface includes 24 scatterers, whose centers describe two circles with the radius of r_1 and r_2.

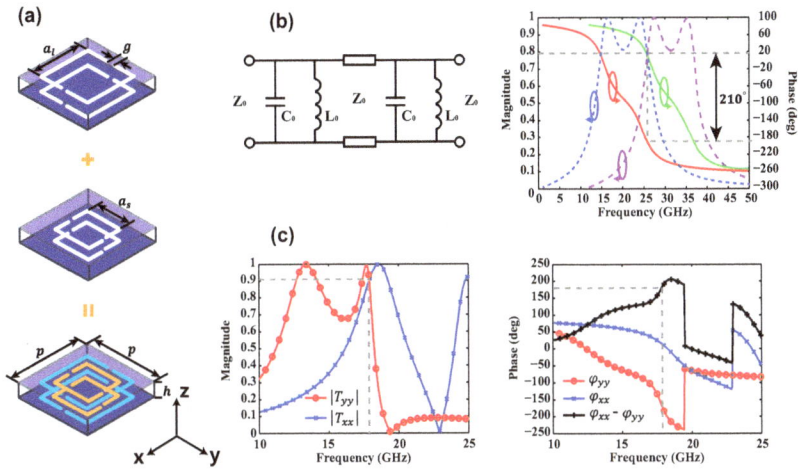

Figure 4. Schematic and the response of the proposed unit cell: (**a**) Schematic; (**b**) Equivalent circuit model of one pair of complementary split-ring resonators (CSRRs) and simulated S_{21}. The purple and green curves are obtained by the translation of the original blue (magnitude) and red (phase) curves. The distance between the two layers is $h = 1.25$ mm. Characteristic impedance of free space is $Z_0 = 377\ \Omega$. The capacitance and inductance are $C_0 = 0.09$ pF and $L_0 = 1.03$ nH; (**c**) Full-wave simulation results. The period of the unit cell is 7×7 mm^2. Side lengths of the two types of square CSRRs are $a_l = 5.2$ mm and $a_s = 3.9$ mm. The length of the gap is $g = 0.2$ mm. The width of the slots is $t = 0.2$ mm. Reproduced with permission from [85], Copyright IEEE, 2017.

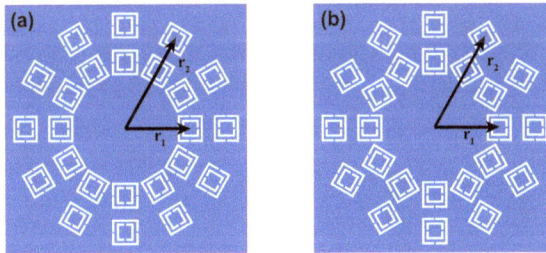

Figure 5. Geometric structure of the metasurface. Topological charge is (**a**) $q = 1$; (**b**) $q = 2$. The radius of the inner ring is $r_1 = 14$ mm and that of the outer ring is $r_2 = 21$ mm. Reproduced with permission from [85], Copyright IEEE, 2017.

The EM responses from the metasurfaces are calculated by an equivalent dipole model and CST MWS. In the equivalent dipole model, each unit cell is considered to be two orthogonal magnetic dipole sources with the orientations aligned with the long side of the CSRRs. Therefore, the EM response from the whole metasurface is considered to be a sum of the response from magnetic dipoles with varying orientations. This approach is useful to simulate a metasurface comprising a great number of scatterers that can be treated as point sources with arbitrary strengths, locations and orientations. The calculated field distributions are shown in Figure 6. The results verify the successful generation of OAM. The divergence between the results from the dipole model and the full-wave simulation comes from the coupling between unit cells.

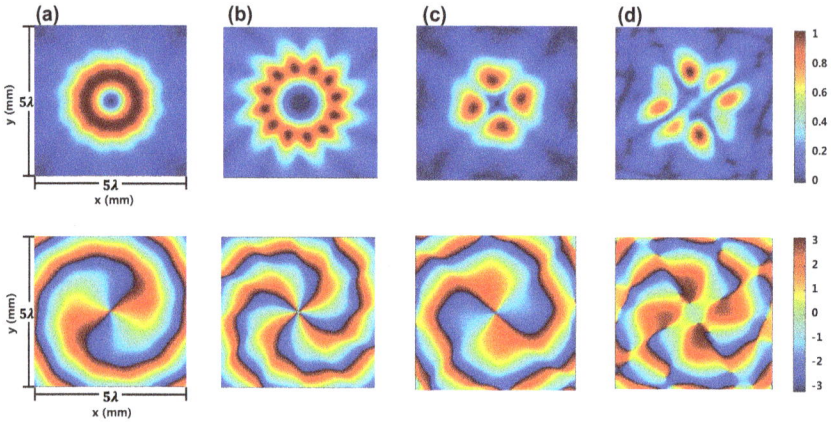

Figure 6. Amplitude and phase distributions of the cross-circularly polarized component of electric field at a transverse plane of $z = 10$ mm calculated from (**a**) the equivalent dipole model with the aperture in Figure 5a; (**b**) the equivalent dipole model with the aperture in Figure 5b; (**c**) the full-wave simulation with the aperture in Figure 5a; (**d**) the full-wave simulation with the aperture in Figure 5b. Reproduced with permission from [85], Copyright IEEE, 2017.

3.3. Metasurface Fork Gratings

The transmission function for a diffraction grating for OAM generation is written by

$$t(r, \phi) = \sum_m A_m e^{j(l_m \phi + k_{xm} x + k_{ym} y)}, \tag{5}$$

where r is the radial position, ϕ is the azimuthal position, A_m is the weight of the mth beam, l_m is the corresponding OAM index, and k_{xm}, k_{ym} are the transverse wave numbers of the mth beam. It can be considered as a hologram resulting from the interference of a plane wave and a OAM wave.

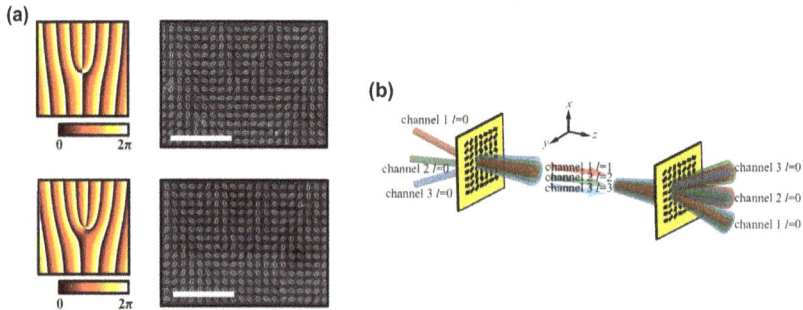

Figure 7. (**a**) Phase distribution and the scanning electron images of metasurface fork gratings with topological charge of $q = 2, 3$; The plasmonic metasurfaces are fabricated on an 80-nm thick aluminum thin film by using focus ion beam method, consists of spatially variant nanoslits with a size of ~50 nm by 210 nm. Scale bar: 3 μm. Reproduced with permission from [92], Copyright WILEY-VCH Verlag GmbH & Co. KGaA, Weinheim, 2016; (**b**) Schematics of off-axis incidence multi-OAM multiplexer and off-axis multi-OAM demultiplexer. Reproduced with permission from [86], Copyright WILEY-VCH Verlag GmbH & Co. KGaA, Weinheim, 2016.

The transmission function is calculated based on the design requirement and the phase information is extracted and reconstructed using geometric-phase metasurfaces. We show two designs in Figure 7 [86,92]. They are formed by nanoslits with varying orientations. Figure 7a shows the phased holograms and the corresponding distributions of nanoslits for a single OAM beam generation. During the fabrication process, first, an 80-nm-think aluminum film is deposited on glass through thermal evaporation process. Then, the nanoslits array is fabricated through focus-ion-beam method. Figure 7b illustrates a general map of three-OAM-channel multiplexing and demultiplexing using metasurfaces. Two identical metasurfaces are used for the multiple OAM-beam generation and separation, respectively. In both cases of Figure 7, the metasurfaces are illuminated by a circularly polarized wave and the cross-circularly polarized component carries OAM.

3.4. Metasurface for OAM-Carrying Vector Beams Generation

In the previous discussion, the metasurfaces generate circularly polarized waves with OAM. In the following, we will review several pieces of research work dealing with vector fields carrying OAM [93,94]. The polarization of vector fields is represent by $\alpha(\phi) = m\phi + \alpha_0$, where m is the polarization order and α_0 is the initial polarization. To generate a vector field, a linear polarizer shown in Figure 8 is used. It composes of rectangular apertures and the transmitted wave is linearly polarized along the direction that is vertical to the long axis of the aperture. When the linear polarizers are oriented to different directions according to their locations, the polarizations are varied locally. The aperture with $\alpha = 0$ is modelled by Jones matrix:

$$\mathbf{J}^l_{lin\_pol} = \begin{pmatrix} 1 & 0 \\ 0 & 0 \end{pmatrix}. \tag{6}$$

Then, the Jones matrix for a rotated aperture is written as

$$\mathbf{J}^l_{lin\_pol}(\alpha) = \begin{pmatrix} \cos^2(\alpha) & \sin(\alpha)\cos(\alpha) \\ \sin(\alpha)\cos(\alpha) & \sin^2(\alpha) \end{pmatrix}. \tag{7}$$

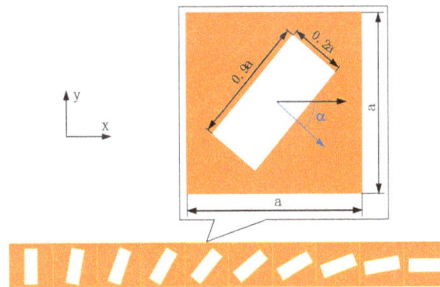

Figure 8. Geometry of a one-dimensional inhomogeneous anisotropic metamaterial composed of 10 rectangular holes with the orientation changed stepwisely from 0 to $\pi/2$. The inset is the geometry of the unit cell that is a square metal slab punched into a rectangular hole. Reproduced with permission from [93], Copyright The Optical Society, 2012.

If the incident wave is circularly polarized, i.e., $E^{\pm}_{in} = \frac{1}{\sqrt{2}}[1 \pm i]^T$ (plus sign for LHCP and minus sign for RHCP), the transmitted field is expressed by

$$E^+_{out} = \mathbf{J}^l_{lin\_pol}E^+_{in} = \frac{1}{\sqrt{2}}e^{i\alpha}\begin{pmatrix} \cos(\alpha) \\ \sin(\alpha) \end{pmatrix} = \frac{1}{2}\begin{pmatrix} 1 \\ i \end{pmatrix} + \frac{1}{2}e^{2i\alpha}\begin{pmatrix} 1 \\ -i \end{pmatrix}. \tag{8}$$

$$E_{out}^- = \mathbf{J}_{lin\_pol}^l E_{in}^- = \frac{1}{\sqrt{2}} e^{-i\alpha} \begin{pmatrix} \cos(\alpha) \\ \sin(\alpha) \end{pmatrix} = \frac{1}{2} e^{-2i\alpha} \begin{pmatrix} 1 \\ i \end{pmatrix} + \frac{1}{2} \begin{pmatrix} 1 \\ -i \end{pmatrix}. \tag{9}$$

From (8) and (9), we can notice that the output wave is linearly polarized with a geometric phase of $e^{\pm i\alpha}$. The linearly polarized wave can be decomposed into two circularly polarized waves. The co-circular polarization does not have the phase term and the cross-circular polarization has the term of $e^{\pm 2i\alpha}$. It should be emphasized that unlike the previous derivation that the conversion efficiency can reach 1, the transmitted power after the linear polarizer is only half of the power in the incident wave and each transmitted circularly polarized component takes half of the total transmitted power.

In Figure 9, two rings of slots are put on a gold film with thickness of 200 nm. The rotation angle of the slots α satisfies $\alpha(\phi) = l\phi + \alpha_0$. Thus, under the excitation of RCP light, according to Equation (9), the spatially variant factor in E_{out} is $e^{-i(l\phi+\alpha_0)}$. The generated OAM order is $-l$. The structure can be fabricated using electron-beam lithography [95].

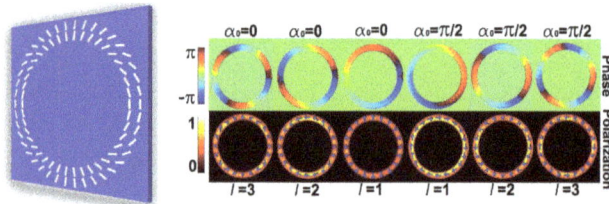

Figure 9. Schematic structure of metamaterials for generating OAM-carrying vector beams and the spatial distributions of phase and polarization of generated OAM-carrying vector beams ($\sigma = -1$, right circularly polarized (RCP) input beam). Reproduced with permission from [94], Copyright The Optical Society, 2013.

3.5. Continuously Shaped Metasurfaces

The scatterers in the previous discussion provide discrete levels of abrupt phase shift by employing the geometric phase concept. In this section, we review several prototypes with continuous or quasi-continuous phase levels [87,88,96]. They generate OAM with high purity.

In Figure 10a, a metasurface is composed of annular apertures with smoothly changed widths. The phase shift comes not only from the geometric phase due to the varying aperture orientation but also from the plasmon retardation phase which is modulated by the aperture width. With the contribution of the two phases, arbitrary OAM orders can be generated. Under the incidence of circularly polarized wave, according to (9), the radially polarized component acquires a geometric phase of $e^{i\phi}$ and the cross-circularly polarized one has a geometric phase of $e^{2i\phi}$. To fabricate the metasurface sample, a 2-nm-thick Cr film and a 450-nm-thick silver film were subsequently deposited on a quartz substrate using magnetron sputtering. The annular apertures are then milled on the silver/Cr film through focused-iron beam lithography. Figure 10b shows the a pattern constructed from catenary-shaped atoms. The inclination angle for a single caternary gradually varies a total of pi from one end of the caternary to the other end. The induced geometric phase doubles the value of inclination value for the cross-circularly polarized output wave. By arranging the catenaries accordingly, OAM beams can be produced. Since this geometric phase does not depend on frequency, the design has a broadband response. The fabrication process of the caternaries is similar to that of the annular apertures in Figure 10a, except that a 120-nm-thick Au film was deposited instead of the 450-nm-thick silver film. The direction of the grating grooves in Figure 10c is designed to satisfy $\theta(r, \phi) = l\phi/2$. It is illuminated by a RCP wave and the transmitted LCP wave carries an OAM of order l. The gratings were created by etching of a 500-μm-thick GaAs wafer through electron cyclotron resonance source (BCl$_3$) to a depth of 2.5-μm.

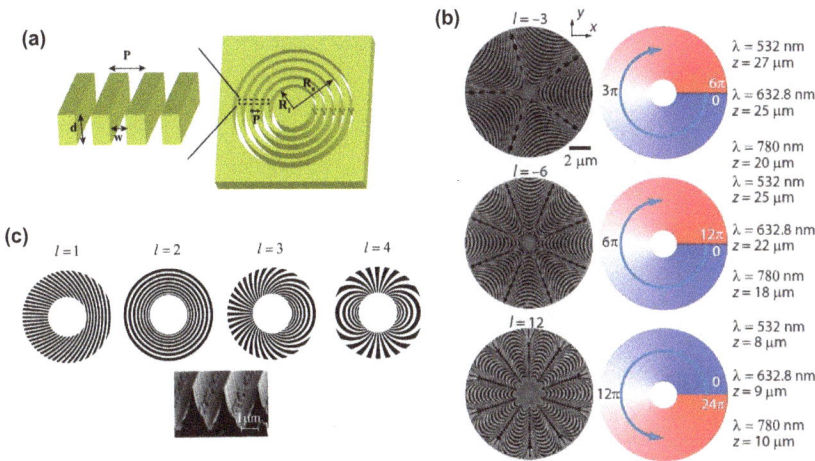

Figure 10. (a) The metasurface is constructed by drilling a silver film with multiple periods of annular rings, whose radius is defined as $R_n = R_1 + (n-1)P$, where n and P denote the number and the period of the apertures. The annular apertures can be taken as two-dimensional extensions of a set of nanoslits with spatially varying orientation. Reproduced with permission from [87], Copyright American Chemical Society, 2016; **(b)** OAM generators based on catenary arrays. The topological charges from up to bottom are -3, -6, and 12 ($s = 1$), respectively. The first column represents the scanning electron microscopy (SEM) images of the fabricated samples. The second column shows the spiral phase profiles. Reproduced with permission from [88], Copyright The American Association for the Advancement of Science, 2015; **(c)** Top, geometry of the subwavelength gratings for four topological charges. Bottom, image of a typical grating profile taken with a scanning-electron microscope. Reproduced with permission from [96], Copyright Optical Society of America, 2002.

3.6. Metasurfaces for Multiple OAM-Beam Generation

Figure 11 depicts two types of geometric-phase metasurfaces for multiple OAM-beam generation [90,97]. The metasurface in Figure 11a generates two collinear OAM beams. The dielectric nanofins forming the topological charges of $q = 2.5$ and $q = 5$ are interleaved each other. The dielectric nanofins consisted of TiO_2 and were fabricated based on atomic layer deposition and electron beam lithography. In Figure 11b, the nanoantennas with different topological charge are interleaved randomly.

4. Holographic Metasurfaces for OAM Detection

We adopt the holographic concept for the detection of multiple OAM-beam using a single metasurface [89]. The detection process is summarized in Figure 12. The metasurface converts the incident wave to multiple waves, only one of which is Gaussian. The radiation direction of the Gaussian wave is distinguishable according to the order of incident OAM. Consequently, by locating the Gaussian wave, the incident OAM can be conveniently determined.

The transmission function of the desired metasurface is given by (5). Then, the far-field response of the metasurface illuminated by an incident wave carrying OAM of order l_0 is calculated by doing the Fourier transform

$$E = F\{E_{in} \cdot t\} = \sum_m A_m F\{E_{OAM(l_m + l_0)}(k_{xm}, k_{ym})\}. \tag{10}$$

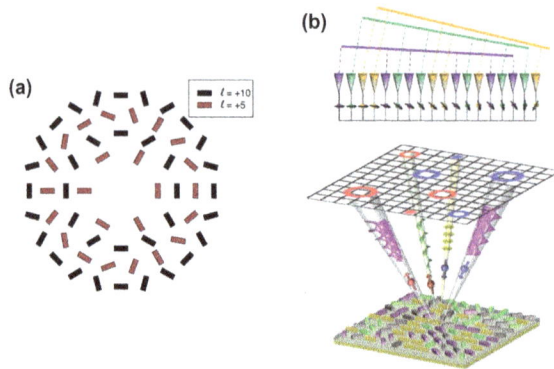

Figure 11. (**a**) Schematic of the nanofins azimuthal distribution in the inner part of metasurface device with interleaved patterns that generate collinear beams having topological charges $|l| = 5$ and $|l| = 10$. The device has a 500 µm diameter and contains more than 700 interleaved radial rows of nanofins. Reproduced with permission from [90], Copyright Optical Society of America, 2017; (**b**) Schematic of shared-aperture concepts using interleaved 1D phased arrays and the schematic far-field intensity distribution of wavefronts with positive (red) and negative (blue) helicities. Reproduced with permission from [97], Copyright The American Association for the Advancement of Science, 2016.

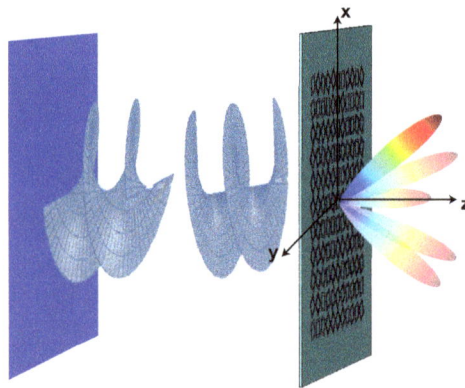

Figure 12. Schematic representation of multiple OAM-beam detection by a single metasurface. Reproduced with permission from [89], Copyright IEEE, 2017.

Multiple waves with the OAM of order $l_m + l_0$ at the k-space position (k_{xm}, k_{ym}) are observed. When $l_M + l_0 = 0$, the beam is Gaussian and its beam axis is at (k_{xM}, k_{yM}). It is known that OAM wave has a singularity at its beam axis. Therefore, by examining the field intensity at the positions of (k_{xm}, k_{ym}), we can identify the gaussian beam, i.e., identify M. Then the incident OAM order l_0 can be determined.

As a proof of concept, a five-beam case with $l_m = 2, 1, 0, -1, -2$ at the directions of $\theta = 40°$ and $\phi = 90°, 18°, 306°, 234°, 162°$ is demonstrated. The calculated transmittance $t(r, \phi)$ is implemented using the unit cell as shown in Figure 4a. Full-wave simulated radiation patterns are shown in Figure 13. It can be seen that the maximum radiation direction (axis of the gaussian beam) when $l_0 = -2, -1, 0, 1, 2$ is at $\theta = 40°$ and $\phi = 90°, 18°, 306°, 234°, 162°$, respectively, which is as expected.

A modified transmission function is proposed to lower the side-lobe level:

$$t_{mod}(r,\phi) = \sum_m e^{j(l_m\phi + k_{xm}x + k_{ym}y + \alpha_m)}. \tag{11}$$

We see an additional phase term $e^{j\alpha_m}$ in (11). This term rotates the mth beam and changes the interference status between the beam with other four beams. Therefore, by setting a proper value for α_m, we can weaken the constructive interference between adjacent beams, which is the main reason for the high side lobes. The optimal solution for α_m is $\alpha_1 = 1.0472$, $\alpha_2 = 1.0472$, $\alpha_3 = 2.0944$, $\alpha_4 = 2.7925$, $\alpha_5 = 4.5379$. The comparison results after optimization are shown in Figure 13f. When $l_0 = \pm1, \pm2$, the field intensity at a desired location is increased. A general trend of lowered side-lobe level can also be observed. It can be noted that the intensity becomes lower for $l_0 = 0$. One may break the constructive interference between two adjacent beams but result in an enhanced interference between one of the adjacent beam with the other adjacent beam on the other side. Hence, it is not possible to achieve improvements for all the five incident cases, but there has to be a trade off. Overall, we can observe suppressed side lobes and increased field intensities at desired locations.

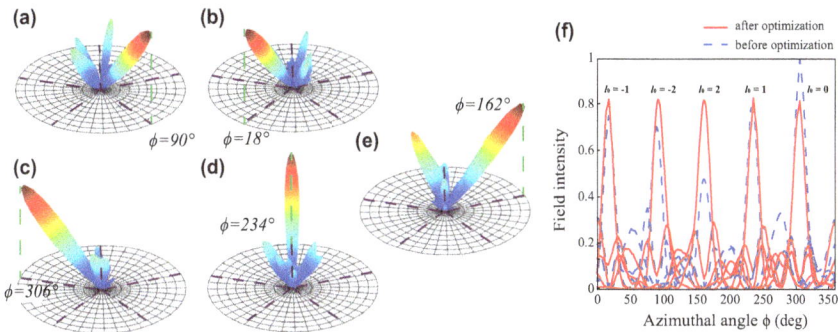

Figure 13. Full-wave simulated far-field power patterns when the incident wave carries OAM of order (**a**) −2; (**b**) −1; (**c**) 0; (**d**) 1; (**e**) 2. (**f**) Original and optimized far-field power patterns at $\theta = 40°$ for the five cases (**a**–**e**). Reproduced with permission from [89], Copyright IEEE, 2017.

5. Conclusions

In summary, we reviewed the research work on the geometric-phase based metasurfaces for OAM generation and detection. The metasurfaces achieve wavefront manipulation by the spin-induced geometric phase and show high flexibilities. Most importantly, they can be multifunctional. Besides the OAM generation and detection, they are designed for realizing beam multiplexing and demultiplexing and manipulating polarization. Therefore, the geometric-phase metasurfaces are promising candidates for practical applications of OAM beams.

Acknowledgments: This work was supported in part by the Research Grants Council of Hong Kong (GRF 716713, GRF 17207114, and GRF 17210815), NSFC 61271158, Hong Kong UGC AoE/P04/08, AOARD FA2386-17-1-0010, Hong Kong ITP/045/14LP, and Hundred Talents Program of Zhejiang University (No. 188020*194231701/208).

Author Contributions: Menglin L. N. Chen drafted the manusciprt. Li Jun Jiang and Wei. E. I. Sha revised and finalized the manuscript.

Conflicts of Interest: The authors declare no conflict of interest.

Abbreviations

The following abbreviations are used in this manuscript:

OAM	Orbital angular momentum
EM	Electromagnetic
AM	Angular momentum
SAM	Spin angular momentum
LHCP	Left-handed circular polarization
RHCP	Right-handed circular polarization
LG	Laguerre-Gaussian
CGH	Computer generated hologram
SPP	Spiral phase plates
FSS	Frequency selective surface
RCP	Right circularly polarized
LCP	Left circularly polarized
SRR	Split-ring resonators
PEC	Perfect electric conductor
PMC	Perfect magnetic conductor
PCB	Printed circuit board
CSRR	Complementary split-ring resonators
LC	Inductor-capacitor

References

1. Bliokh, K.Y.; Bekshaev, A.Y.; Nori, F. Dual electromagnetism: Helicity, spin, momentum and angular momentum. *New J. Phys.* **2013**, *15*, 033126.
2. Gibson, G.; Courtial, J.; Padgett, M.J.; Vasnetsov, M.; Pas'ko, V.; Barnett, S.M.; Franke-Arnold, S. Free-space information transfer using light beams carrying orbital angular momentum. *Opt. Express* **2004**, *12*, 5448–5456.
3. Ren, Y.X.; Wang, Z.; Liao, P.C.; Li, L.; Xie, G.D.; Huang, H.; Zhao, Z.; Yan, Y.; Ahmed, N.; Willner, A.; et al. Experimental characterization of a 400 Gbit/s orbital angular momentum multiplexed free-space optical link over 120 m. *Opt. Lett.* **2016**, *41*, 622–625.
4. Willner, A.E.; Huang, H.; Yan, Y.; Ren, Y.; Ahmed, N.; Xie, G.; Bao, C.; Li, L.; Cao, Y.; Zhao, Z.; et al. Optical communications using orbital angular momentum beams. *Adv. Opt. Photonics* **2015**, *7*, 66–106.
5. Wang, J.; Yang, J.Y.; Fazal, I.M.; Ahmed, N.; Yan, Y.; Huang, H.; Ren, Y.X.; Yue, Y.; Dolinar, S.; Tur, M.; et al. Terabit free-space data transmission employing orbital angular momentum multiplexing. *Nat. Photonics* **2012**, *6*, 488–496.
6. Thide, B.; Then, H.; Sjoholm, J.; Palmer, K.; Bergman, J.; Carozzi, T.D.; Istomin, Y.N.; Ibragimov, N.H.; Khamitova, R. Utilization of photon orbital angular momentum in the low-frequency radio domain. *Phys. Rev. Lett.* **2007**, *99*, 087701.
7. Mahmouli, F.E.; Walker, S.D. 4-Gbps uncompressed video transmission over a 60-ghz orbital angular momentum wireless channel. *IEEE Wirel. Commun. Lett.* **2013**, *2*, 223–226.
8. Yan, Y.; Xie, G.D.; Lavery, M.P.J.; Huang, H.; Ahmed, N.; Bao, C.J.; Ren, Y.X.; Cao, Y.W.; Li, L.; Zhao, Z.; et al. High-capacity millimetre-wave communications with orbital angular momentum multiplexing. *Nat. Commun.* **2014**, *5*, doi:10.1038/ncomms5876.
9. Hui, X.N.; Zheng, S.L.; Chen, Y.L.; Hu, Y.P.; Jin, X.F.; Chi, H.; Zhang, X.M. Multiplexed Millimeter Wave Communication with Dual Orbital Angular Momentum (OAM) Mode Antennas. *Sci. Rep.* **2015**, *5*, 10148.
10. Bozinovic, N.; Yue, Y.; Ren, Y.X.; Tur, M.; Kristensen, P.; Huang, H.; Willner, A.E.; Ramachandran, S. Terabit-Scale Orbital Angular Momentum Mode Division Multiplexing in Fibers. *Science* **2013**, *340*, 1545–1548.
11. Andersson, M.; Berglind, E.; Björk, G. Orbital angular momentum modes do not increase the channel capacity in communication links. *New J. Phys.* **2015**, *17*, 043040.
12. Xie, G.D.; Li, L.; Ren, Y.X.; Huang, H.; Yan, Y.; Ahmed, N.; Zhao, Z.; Lavery, M.P.J.; Ashrafi, N.; Ashrafi, S.; et al. Performance metrics and design considerations for a free-space optical orbital-angular-momentum-multiplexed communication link. *Optica* **2015**, *2*, 357–365.

13. Paterson, C. Atmospheric turbulence and orbital angular momentum of single photons for optical communication. *Phys. Rev. Lett.* **2005**, *94*, 153901.

14. Zhao, S.M.; Leach, J.; Gong, L.Y.; Ding, J.; Zheng, B.Y. Aberration corrections for free-space optical communications in atmosphere turbulence using orbital angular momentum states. *Opt. Express* **2012**, *20*, 452–461.

15. Oldoni, M.; Spinello, F.; Mari, E.; Parisi, G.; Someda, C.G.; Tamburini, F.; Romanato, F.; Ravanelli, R.A.; Coassini, P.; Thide, B. Space-Division Demultiplexing in Orbital-Angular-Momentum-Based MIMO Radio Systems. *IEEE Trans. Antennas Propag.* **2015**, *63*, 4582–4587.

16. Tamburini, F.; Anzolin, G.; Umbriaco, G.; Bianchini, A.; Barbieri, C. Overcoming the Rayleigh criterion limit with optical vortices. *Phys. Rev. Lett.* **2006**, *97*, 163903.

17. Liu, K.; Cheng, Y.Q.; Yang, Z.C.; Wang, H.Q.; Qin, Y.L.; Li, X. Orbital-Angular-Momentum-Based Electromagnetic Vortex Imaging. *IEEE Antennas Wirel. Propag. Lett.* **2015**, *14*, 711–714.

18. Grier, D.G. A revolution in optical manipulation. *Nature* **2003**, *424*, 810–816.

19. He, H.; Friese, M.E.J.; Heckenberg, N.R.; Rubinszteindunlop, H. Direct Observation of Transfer of Angular-Momentum to Absorptive Particles from a Laser-Beam with a Phase Singularity. *Phys. Rev. Lett.* **1995**, *75*, 826–829.

20. Simpson, N.B.; Dholakia, K.; Allen, L.; Padgett, M.J. Mechanical equivalence of spin and orbital angular momentum of light: an optical spanner. *Opt. Lett.* **1997**, *22*, 52–54.

21. Lavery, M.P.J.; Speirits, F.C.; Barnett, S.M.; Padgett, M.J. Detection of a Spinning Object Using Light's Orbital Angular Momentum. *Science* **2013**, *341*, 537–540.

22. Dada, A.C.; Leach, J.; Buller, G.S.; Padgett, M.J.; Andersson, E. Experimental high-dimensional two-photon entanglement and violations of generalized Bell inequalities. *Nat. Phys.* **2011**, *7*, 677–680.

23. Leach, J.; Jack, B.; Romero, J.; Jha, A.K.; Yao, A.M.; Franke-Arnold, S.; Ireland, D.G.; Boyd, R.W.; Barnett, S.M.; Padgett, M.J. Quantum Correlations in Optical Angle-Orbital Angular Momentum Variables. *Science* **2010**, *329*, 662–665.

24. Ren, H.R.; Li, X.P.; Zhang, Q.M.; Gu, M. On-chip noninterference angular momentum multiplexing of broadband light. *Science* **2016**, *352*, 805–809.

25. Mei, S.T.; Huang, K.; Liu, H.; Qin, F.; Mehmood, M.Q.; Xu, Z.J.; Hong, M.H.; Zhang, D.H.; Teng, J.H.; Danner, A.; et al. On-chip discrimination of orbital angular momentum of light with plasmonic nanoslits. *Nanoscale* **2016**, *8*, 2227–2233.

26. Molina-Terriza, G.; Torres, J.P.; Torner, L. Twisted photons. *Nat. Phys.* **2007**, *3*, 305–310.

27. Vaziri, A.; Weihs, G.; Zeilinger, A. Superpositions of the orbital angular momentum for applications in quantum experiments. *J. Opt. B* **2002**, *4*, S47–S51.

28. Allen, L.; Beijersbergen, M.W.; Spreeuw, R.J.C.; Woerdman, J.P. Orbital Angular-Momentum of Light and the Transformation of Laguerre-Gaussian Laser Modes. *Phys. Rev. A* **1992**, *45*, 8185–8189.

29. Bazhenov, V.Y.; Vasnetsov, M.V.; Soskin, M.S. Laser-Beams with Screw Dislocations in Their Wave-Fronts. *JETP Lett.* **1990**, *52*, 429–431.

30. Beijersbergen, M.W.; Allen, L.; Vanderveen, H.E.L.O.; Woerdman, J.P. Astigmatic Laser Mode Converters and Transfer of Orbital Angular-Momentum. *Opt. Commun.* **1993**, *96*, 123–132.

31. Beijersbergen, M.; Coerwinkel, R.; Kristensen, M.; Woerdman, J. Helical-wavefront laser beams produced with a spiral phaseplate. *Opt. Commun.* **1994**, *112*, 321–327.

32. Marrucci, L.; Manzo, C.; Paparo, D. Optical spin-to-orbital angular momentum conversion in inhomogeneous anisotropic media. *Phys. Rev. Lett.* **2006**, *96*, 163905.

33. Mohammadi, S.M.; Daldorff, L.K.S.; Bergman, J.E.S.; Karlsson, R.L.; Thide, B.; Forozesh, K.; Carozzi, T.D.; Isham, B. Orbital Angular Momentum in Radio-A System Study. *IEEE Trans. Antennas Propag.* **2010**, *58*, 565–572.

34. Zheng, S.L.; Hui, X.N.; Jin, X.F.; Chi, H.; Zhang, X.M. Transmission Characteristics of a Twisted Radio Wave Based on Circular Traveling-Wave Antenna. *IEEE Trans. Antennas Propag.* **2015**, *63*, 1530–1536.

35. Barbuto, M.; Trotta, F.; Bilotti, F.; Toscano, A. Circular Polarized Patch Antenna Generating Orbital Angular Momentum. *Prog. Electromagn. Res.* **2014**, *148*, 23–30.

36. Yu, N.F.; Capasso, F. Flat optics with designer metasurfaces. *Nat. Mater.* **2014**, *13*, 139–150.

37. Yu, N.F.; Genevet, P.; Kats, M.A.; Aieta, F.; Tetienne, J.P.; Capasso, F.; Gaburro, Z. Light Propagation with Phase Discontinuities: Generalized Laws of Reflection and Refraction. *Science* **2011**, *334*, 333–337.

38. Genevet, P.; Yu, N.F.; Aieta, F.; Lin, J.; Kats, M.A.; Blanchard, R.; Scully, M.O.; Gaburro, Z.; Capasso, F. Ultra-thin plasmonic optical vortex plate based on phase discontinuities. *Appl. Phys. Lett.* **2012**, *100*, 013101.

39. Yu, N.F.; Genevet, P.; Aieta, F.; Kats, M.A.; Blanchard, R.; Aoust, G.; Tetienne, J.P.; Gaburro, Z.; Capasso, F. Flat optics: Controlling wavefronts with optical antenna metasurfaces. *IEEE J. Sel. Top. Quantum Electron.* **2013**, *19*, doi:10.1109/JSTQE.2013.2241399.

40. Munk, B. *Frequency Selective Surfaces: Theory and Design*; John Wiley: New York, NY, USA, 2000.

41. Kou, N.; Yu, S.X.; Li, L. Generation of high-order Bessel vortex beam carrying orbital angular momentum using multilayer amplitude-phase-modulated surfaces in radiofrequency domain. *Appl. Phys. Express* **2017**, *10*, 016701.

42. Pfeiffer, C.; Grbic, A. Metamaterial Huygens' Surfaces: Tailoring Wave Fronts with Reflectionless Sheets. *Phys. Rev. Lett.* **2013**, *110*, 197401.

43. Niv, A.; Biener, G.; Kleiner, V.; Hasman, E. Spiral phase elements obtained by use of discrete space-variant subwavelength gratings. *Opt. Commun.* **2005**, *251*, 306–314.

44. Turnbull, G.A.; Robertson, D.A.; Smith, G.M.; Allen, L.; Padgett, M.J. The generation of free-space Laguerre-Gaussian modes at millimetre-wave frequencies by use of a spiral phaseplate. *Opt. Commun.* **1996**, *127*, 183–188.

45. Kotlyar, V.V.; Almazov, A.A.; Khonina, S.N.; Soifer, V.A.; Elfstrom, H.; Turunen, J. Generation of phase singularity through diffracting a plane or Gaussian beam by a spiral phase plate. *J. Opt. Soc. Am. A* **2005**, *22*, 849–861.

46. Hui, X.N.; Zheng, S.L.; Hu, Y.P.; Xu, C.; Jin, X.F.; Chi, H.; Zhang, X.M. Ultralow Reflectivity Spiral Phase Plate for Generation of Millimeter-wave OAM Beam. *IEEE Antennas Wirel. Propag. Lett.* **2015**, *14*, 966–969.

47. Cheng, L.; Hong, W.; Hao, Z.C. Generation of Electromagnetic Waves with Arbitrary Orbital Angular Momentum Modes. *Sci. Rep.* **2014**, *4*, doi:10.1038/srep04814.

48. Sun, J.B.; Wang, X.; Xu, T.B.Y.; Kudyshev, Z.A.; Cartwright, A.N.; Litchinitser, N.M. Spinning Light on the Nanoscale. *Nano Lett.* **2014**, *14*, 2726–2729.

49. Carpentier, A.V.; Michinel, H.; Salgueiro, J.R.; Olivieri, D. Making optical vortices with computer-generated holograms. *Am. J. Phys.* **2008**, *76*, 916–921.

50. Heckenberg, N.R.; Mcduff, R.; Smith, C.P.; White, A.G. Generation of Optical-Phase Singularities by Computer-Generated Holograms. *Opt. Lett.* **1992**, *17*, 221–223.

51. Arlt, J.; Dholakia, K.; Allen, L.; Padgett, M.J. The production of multiringed Laguerre-Gaussian modes by computer-generated holograms. *J. Mod. Opt.* **1998**, *45*, 1231–1237.

52. Moreno, I.; Davis, J.A.; Pascoguin, B.M.L.; Mitry, M.J.; Cottrell, D.M. Vortex sensing diffraction gratings. *Opt. Lett.* **2009**, *34*, 2927–2929.

53. Liu, K.; Liu, H.Y.; Qin, Y.L.; Cheng, Y.Q.; Wang, S.N.; Li, X.; Wang, H.Q. Generation of OAM Beams Using Phased Array in the Microwave Band. *IEEE Trans. Antennas Propag.* **2016**, *64*, 3850–3857.

54. Comite, D.; Valerio, G.; Albani, M.; Galli, A.; Casaletti, M.; Ettorre, M. Exciting Vorticity through Higher Order Bessel Beams with a Radial-Line Slot-Array Antenna. *IEEE Trans. Antennas Propag.* **2017**, *65*, 2123–2128.

55. Xu, B.J.; Wu, C.; Wei, Z.Y.; Fan, Y.C.; Li, H.Q. Generating an orbital-angular-momentum beam with a metasurface of gradient reflective phase. *Opt. Mater. Express* **2016**, *6*, 3940–3945.

56. Yu, S.X.; Li, L.; Shi, G.M.; Zhu, C.; Shi, Y. Generating multiple orbital angular momentum vortex beams using a metasurface in radio frequency domain. *Appl. Phys. Lett.* **2016**, *108*, 241901.

57. Wang, L.Y.; Shi, H.Y.; Zhu, S.T.; Li, J.X.; Zhang, A.X.; Li, L.M. Generation of multiple modes microwave vortex beams using tunable metasurfacee. In Proceedings of the 7th IEEE International Symposium on Microwave, Antenna, Propagation and EMC Technologies, Xi'an, China, 24–27 October 2017; pp. 379–381.

58. Maccalli, S.; Pisano, G.; Colafrancesco, S.; Maffei, B.; Ng, M.W.R.; Gray, M. Q-plate for millimeter-wave orbital angular momentum manipulation. *Appl. Opt.* **2013**, *52*, 635–639.

59. Cardano, F.; Karimi, E.; Slussarenko, S.; Marrucci, L.; de Lisio, C.; Santamato, E. Polarization pattern of vector vortex beams generated by q-plates with different topological charges. *Appl. Opt.* **2012**, *51*, C1–C6.

60. Bliokh, K.Y.; Rodriguez-Fortuno, F.J.; Nori, F.; Zayats, A.V. Spin-orbit interactions of light. *Nat. Photonics* **2015**, *9*, 796–808.

61. Karimi, E.; Piccirillo, B.; Nagali, E.; Marrucci, L.; Santamato, E. Efficient generation and sorting of orbital angular momentum eigenmodes of light by thermally tuned q-plates. *Appl. Phys. Lett.* **2009**, *94*, 231124.

62. Piccirillo, B.; D'Ambrosio, V.; Slussarenko, S.; Marrucci, L.; Santamato, E. Photon spin-to-orbital angular momentum conversion via an electrically tunable q-plate. *Appl. Phys. Lett.* **2010**, *97*, 241104.

63. Kang, M.; Feng, T.H.; Wang, H.T.; Li, J.S. Wave front engineering from an array of thin aperture antennas. *Opt. Express* **2012**, *20*, 15882–15890.

64. Xu, H.X.; Liu, H.; Ling, X.; Sun, Y.; Yuan, F. Broadband Vortex Beam Generation Using Multimode Pancharatnam-Berry Metasurface. *IEEE Trans. Antennas Propag.* **2017**, *65*, 7378–7382.

65. Karimi, E.; Schulz, S.A.; De Leon, I.; Qassim, H.; Upham, J.; Boyd, R.W. Generating optical orbital angular momentum at visible wavelengths using a plasmonic metasurface. *Light Sci. Appl.* **2014**, *3*, e167.

66. Tan, Q.L.; Guo, Q.H.; Liu, H.C.; Huang, G.; Zhang, S. Controlling the plasmonic orbital angular momentum by combining the geometric and dynamic phases. *Nanoscale* **2017**, *9*, 4944–4949.

67. Li, G.X.; Kang, M.; Chen, S.M.; Zhang, S.; Pun, E.Y.B.; Cheah, K.W.; Li, J.S. Spin-enabled plasmonic metasurfaces for manipulating orbital angular momentum of light. *Nano Lett.* **2013**, *13*, 4148–4151.

68. Mohammadi, S.M.; Daldorff, L.K.S.; Forozesh, K.; Thide, B.; Bergman, J.E.S.; Isham, B.; Karlsson, R.; Carozzi, T.D. Orbital angular momentum in radio: Measurement methods. *Radio Sci.* **2010**, *45*.

69. Schulze, C.; Dudley, A.; Flamm, D.; Duparre, M.; Forbes, A. Measurement of the orbital angular momentum density of light by modal decomposition. *New J. Phys.* **2013**, *15*, 073025.

70. Hui, X.N.; Zheng, S.L.; Zhang, W.T.; Jin, X.F.; Chi, H.; Zhang, X.M. Local topological charge analysis of electromagnetic vortex beam based on empirical mode decomposition. *Opt. Express* **2016**, *24*, 5423–5430.

71. Zhang, C.; Lu, M. Detecting the orbital angular momentum of electro-magnetic waves using virtual rotational antenna. *Sci. Rep.* **2017**, *7*, 4585.

72. Courtial, J.; Dholakia, K.; Robertson, D.A.; Allen, L.; Padgett, M.J. Measurement of the rotational frequency shift imparted to a rotating light beam possessing orbital angular momentum. *Phys. Rev. Lett.* **1998**, *80*, 3217–3219.

73. Allen, L.; Babiker, M.; Power, W.L. Azimuthal Doppler-Shift in Light-Beams with Orbital Angular-Momentum. *Opt. Commun.* **1994**, *112*, 141–144.

74. Genevet, P.; Lin, J.; Kats, M.A.; Capasso, F. Holographic detection of the orbital angular momentum of light with plasmonic photodiodes. *Nat. Commun.* **2012**, *3*, 1278.

75. Tamburini, F.; Mari, E.; Sponselli, A.; Thide, B.; Bianchini, A.; Romanato, F. Encoding many channels on the same frequency through radio vorticity: first experimental test. *New J. Phys.* **2012**, *14*, 033001.

76. Mair, A.; Vaziri, A.; Weihs, G.; Zeilinger, A. Entanglement of the orbital angular momentum states of photons. *Nature* **2001**, *412*, 313–316.

77. Luo, W.J.; Xiao, S.Y.; He, Q.; Sun, S.L.; Zhou, L. Photonic Spin Hall Effect with Nearly Efficiency. *Adv. Opt. Mater.* **2015**, *3*, 1102–1108.

78. Georgi, P.; Schlickriede, C.; Li, G.X.; Zhang, S.; Zentgraf, T. Rotational Doppler shift induced by spin-orbit coupling of light at spinning metasurfaces. *Optica* **2017**, *4*, 1000–1005.

79. Bouchard, F.; De Leon, I.; Schulz, S.A.; Upham, J.; Karimi, E.; Boyd, R.W. Optical spin-to-orbital angular momentum conversion in ultra-thin metasurfaces with arbitrary topological charges. *Appl. Phys. Lett.* **2014**, *105*, 101905.

80. Wang, W.; Li, Y.; Guo, Z.Y.; Li, R.Z.; Zhang, J.R.; Zhang, A.J.; Qu, S.L. Ultra-thin optical vortex phase plate based on the metasurface and the angular momentum transformation. *J. Opt.* **2015**, *17*, 045102.

81. Jin, J.J.; Luo, J.; Zhang, X.H.; Gao, H.; Li, X.; Pu, M.B.; Gao, P.; Zhao, Z.Y.; Luo, X.G. Generation and detection of orbital angular momentum via metasurface. *Sci. Rep.* **2016**, *6*, doi:10.1038/srep24286.

82. Zelenchuk, D.; Fusco, V. Split-Ring FSS Spiral Phase Plate. *IEEE Antennas Wirel. Propag. Lett.* **2013**, *12*, 284–287.

83. Tan, Y.H.; Li, L.L.; Ruan, H.X. An Efficient Approach to Generate Microwave Vector-Vortex Fields Based on Metasurface. *Microw. Opt. Technol. Lett.* **2015**, *57*, 1708–1713.

84. Chen, M.L.N.; Jiang, L.J.; Sha, W.E.I. Artificial perfect electric conductor-perfect magnetic conductor anisotropic metasurface for generating orbital angular momentum of microwave with nearly perfect conversion efficiency. *J. Appl. Phys.* **2016**, *119*, 064506.

85. Chen, M.L.L.N.; Jiang, L.J.; Sha, W.E.I. Ultrathin Complementary Metasurface for Orbital Angular Momentum Generation at Microwave Frequencies. *IEEE Trans. Antennas Propag.* **2017**, *65*, 396–400.

86. Li, Y.; Li, X.; Chen, L.W.; Pu, M.B.; Jin, J.J.; Hong, M.H.; Luo, X.G. Orbital Angular Momentum Multiplexing and Demultiplexing by a Single Metasurface. *Adv. Opt. Mater.* **2017**, *5*, doi:10.1002/adom.201600502.

87. Guo, Y.H.; Pu, M.B.; Zhao, Z.Y.; Wang, Y.Q.; Jin, J.J.; Gao, P.; Li, X.; Ma, X.L.; Luo, X.G. Merging Geometric Phase and Plasmon Retardation Phase in Continuously Shaped Metasurfaces for Arbitrary Orbital Angular Momentum Generation. *ACS Photonics* **2016**, *3*, 2022–2029.

88. Pu, M.B.; Li, X.; Ma, X.L.; Wang, Y.Q.; Zhao, Z.Y.; Wang, C.T.; Hu, C.G.; Gao, P.; Huang, C.; Ren, H.R.; et al. Catenary optics for achromatic generation of perfect optical angular momentum. *Sci. Adv.* **2015**, *1*, e1500396.

89. Chen, M.; Jiang, L.J.; Wei, E. Detection of Orbital Angular Momentum with Metasurface at Microwave Band. *IEEE Antennas Wirel. Propag. Lett.* **2017**, doi:10.1109/LAWP.2017.2777439.

90. Devlin, R.C.; Ambrosio, A.; Wintz, D.; Oscurato, S.L.; Zhu, A.Y.; Khorasaninejad, M.; Oh, J.; Maddalena, P.; Capasso, F. Spin-to-orbital angular momentum conversion in dielectric metasurfaces. *Opt. Express* **2017**, *25*, 377–393.

91. Dang, X.W.; Li, M.K.; Yang, F.; Xu, S.H. Quasi-periodic array modeling using reduced basis from elemental array. *IEEE J. Multiscale Multiphys. Comput. Tech.* **2017**, *2*, 202–208.

92. Chen, S.M.; Cai, Y.; Li, G.X.; Zhang, S.; Cheah, K.W. Geometric metasurface fork gratings for vortex-beam generation and manipulation. *Laser Photonics Rev.* **2016**, *10*, 322–326.

93. Kang, M.; Chen, J.; Wang, X.L.; Wang, H.T. Twisted vector field from an inhomogeneous and anisotropic metamaterial. *J. Opt. Soc. Am. B* **2012**, *29*, 572–576.

94. Zhao, Z.; Wang, J.; Li, S.H.; Willner, A.E. Metamaterials-based broadband generation of orbital angular momentum carrying vector beams. *Opt. Lett.* **2013**, *38*, 932–934.

95. Wang, J.; Du, J. Plasmonic and dielectric metasurfaces: Design, fabrication and applications. *Appl. Sci.* **2016**, *6*, 239.

96. Biener, G.; Niv, A.; Kleiner, V.; Hasman, E. Formation of helical beams by use of Pancharatnam-Berry phase optical elements. *Opt. Lett.* **2002**, *27*, 1875–1877.

97. Maguid, E.; Yulevich, I.; Veksler, D.; Kleiner, V.; Brongersma, M.L.; Hasman, E. Photonic spin-controlled multifunctional shared-aperture antenna array. *Science* **2016**, *352*, 1202–1206.

applied
sciences

MDPI

Review

Metasurface-Based Polarimeters

Fei Ding *, Yiting Chen * and Sergey I. Bozhevolnyi

SDU Nano Optics, University of Southern Denmark, Campusvej 55, DK-5230 Odense, Denmark;
seib@mci.sdu.dk
* Correspondence: feid@mci.sdu.dk (F.D.); yic@mci.sdu.dk (Y.C.); Tel.: +45-9119-5127 (F.D.)

Received: 1 March 2018; Accepted: 4 April 2018; Published: 10 April 2018

Abstract: The state of polarization (SOP) is an inherent property of light that can be used to gain crucial information about the composition and structure of materials interrogated with light. However, the SOP is difficult to experimentally determine since it involves phase information between orthogonal polarization states, and is uncorrelated with the light intensity and frequency, which can be easily determined with photodetectors and spectrometers. Rapid progress on optical gradient metasurfaces has resulted in the development of conceptually new approaches to the SOP characterization. In this paper, we review the fundamentals of and recent developments within metasurface-based polarimeters. Starting by introducing the concepts of generalized Snell's law and Stokes parameters, we explain the Pancharatnam–Berry phase (PB-phase) which is instrumental for differentiating between orthogonal circular polarizations. Then we review the recent progress in metasurface-based polarimeters, including polarimeters, spectropolarimeters, orbital angular momentum (OAM) spectropolarimeters, and photodetector integrated polarimeters. The review is ended with a short conclusion and perspective for future developments.

Keywords: metasurface-based polarimeters; Stokes parameters; spectropolarimeters; orbit angular momentum; photodetector integrated polarimeters

---

1. Introduction

The state of polarization (SOP) is an inherent property of light and carries crucial information about the composition and structure of materials interrogated with light [1]. However, the SOP is rather cumbersome to probe experimentally, owing to the fact that it is uncorrelated with the intensity and frequency of light, thereby resulting in the loss of information on the relative phase between orthogonal vector components in conventional detection schemes. Hence, polarimeters, which enable direct measurement of the SOP, are greatly desired in many areas of science and technology, including astronomy [2], medical diagnostics [3], and remote sensing [4]. Despite all scientific and technological potential, polarimeters are still very challenging to develop as SOP characterization requires conventionally six intensity measurements to determine the Stokes parameters [5]. Typically, the SOP is probed by utilizing a set of discrete polarizers and wave-plates consecutively placed in front of a detector. By measuring the light flux transmitting through these polarization components, the Stokes parameters that uniquely define the SOP can be determined. Polarimeters based on conventional discrete optical components amount to bulky, expensive, and complicated optical systems that are not compatible with the general trend of integration and miniaturization in photonics and plasmonics.

Metasurfaces are thin planar arrays of resonant subwavelength elements arranged in a periodic or aperiodic (even random) manner which modifies boundary conditions for impinging optical waves in order to realize specific wave shaping. In recent years, metasurfaces have attracted progressively increasing attention and have become a rapidly growing field of research, due to their remarkable ability in manipulating electromagnetic (EM) waves, their versatility, and their ease of on-chip fabrication and integration [6–24]. Such metasurfaces can mimic bulk optics since they are capable of engineering the phase front of reflected and/or refracted optical waves at will. Many ultra-compact flat optical components have been accordingly demonstrated, such as beam steerers [6,25–29], surface wave couplers [30–35], focusing lenses [36–43], optical holograms [44–49], and waveplates [50–54].

Metasurfaces, therefore represent an opportunity for polarimetry to overcome the bulky and expensive architectures of conventional volume optics. In this review, we highlight the recent progress in metasurface-based polarimeters during the past few years and attempt to provide our perspective on this specific branch of applications. The rest of this paper is organized in the following sections. In Section 2, we briefly explain generalized Snell's law, followed by the Pancharatnam–Berry phase (PB-phase), and the Stokes parameters. Section 3 is devoted to the metasurface-based polarimeters, including general polarimeters, spectropolarimeters, orbital angular momentum (OAM) spectropolarimeters, and photodetector integrated polarimeters. Finally, we summarize and provide perspective for future developments in Section 4.

2. Fundamentals

2.1. Generalized Snell's Law

The phase discontinuity or abrupt phase shift at the interface between two media was first introduced by Capasso's group in 2011, resulting in a generalized Snell's laws of reflection and refraction [6]. Figure 1 schematically shows a one-dimensional (1D) system used to derive generalized Snell's laws, where the interface between two media consists of an artificial metasurface introducing a position-dependent phase shift $\Phi(x)$ [6]. Considering a plane wave impinging at an angle of θ_i, generalized Snell's laws of reflection and refraction can be written as:

$$\sin(\theta_r)n_i - \sin(\theta_i)n_i = \frac{\lambda_0}{2\pi}\frac{d\Phi}{dx} \tag{1}$$

$$\sin(\theta_t)n_t - \sin(\theta_i)n_i = \frac{\lambda_0}{2\pi}\frac{d\Phi}{dx}, \tag{2}$$

respectively. Here, n_i and n_t are the refractive indexes of the two media, λ_0 is the wavelength in free space, and θ_r and θ_t are the reflected and refracted angles, respectively. From Equations (1) and (2), it is evident that the reflected/refracted beam can have an arbitrary direction, provided that a nonzero phase gradient ($d\Phi/dx$) along the interface is introduced. It should be noted that in the case of $d\Phi/dx = 0$, we recover the usual laws of reflection and refraction that imply the conservation of in-plane wave vectors.

To realize metasurfaces with phase discontinuity, V-shaped antenna—which support symmetrical and antisymmetrical electric dipole resonance—were demonstrated to control the reflection and refraction of linearly polarized (LP) light in the infrared range, governed by generalized Snell's law [6,25]. However, here 2π phase control is achieved with the transmitted/reflected light polarized orthogonal to the incident wave because V-shaped antennae only support electric dipole resonance, which limits the available phase coverage to π due to its Lorentz-like polarizabilities [55,56]. Additionally, the polarization conversion efficiency for such single non-magnetic metasurfaces is only 25% for the lossless case [57,58]. To solve these problems, one can design gap-surface plasmon metasurfaces (GSPMs) [26,27,30,59] or use all-dielectric meta-atoms [40,41], which ensure the full control of the phase space with high efficiency while maintaining the polarization state.

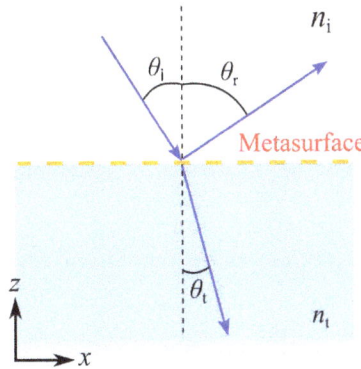

Figure 1. Schematic of a one-dimensional (1D) system with a metasurface positioned at the interface between two media.

2.2. PB-Phase

In the previous examples, phase discontinuity is introduced by varying the meta-atoms' geometric parameters. Another completely different technique, known as PB-phase or geometric phase, achieves full phase control of the cross-polarized light by using anisotropic meta-atoms with identical geometry, but spatially varying the orientations for circularly polarized (CP) light [60–65]. The concept behind PB-phase can be easily revealed by the Jones matrix [66]. In general, the Jones matrix of an anisotropic meta-atom rotated within the *x-y* plane can be written as:

$$M_\theta = R(-\theta) \begin{pmatrix} M_{xx} & 0 \\ 0 & M_{yy} \end{pmatrix} R(\theta), \tag{3}$$

where M_{xx} and M_{yy} are the reflection (or transmission) coefficients for LP light along the two axes of the anisotropic meta-atom, θ is the rotated angle with respect to the *x*-axis, and $R(\theta)$ is the rotation matrix [15]:

$$R(\theta) = \begin{pmatrix} \cos\theta & \sin\theta \\ -\sin\theta & \cos\theta \end{pmatrix}. \tag{4}$$

When the incident light is CP, the reflected or transmitted light can be written as

$$M_\theta \cdot E_0^\pm = \frac{1}{2}\left(M_{xx} + M_{yy}\right) E_0^\pm + \frac{1}{2}\left(M_{xx} - M_{yy}\right) e^{\pm i2\theta} E_0^\mp, \tag{5}$$

where $E_0^\pm$ represents the incident left-handed CP (+, LCP) and right-handed CP (−, RCP) light [15,17]. The first term in Equation (5) represents CP light with the same handedness as the incident wave, while the second term stands for CP light with the opposite handedness and gains an additional PB-phase of $\pm 2\theta$, whose sign depends on the handedness of the incident light. Thereby, 2π phase coverage can be achieved if the meta-atom is rotated from 0 to 180°.

2.3. Stokes Parameters

Before we begin reviewing the recent progress in metasurface-based polarimeters, we briefly introduce the connection between the SOP and the Stokes parameters that are usually measured in experiments [5,67,68]. Consider a monochromatic plane wave that propagates along the *z*-direction, its electric field can be described by the following Jones matrix:

$$\mathbf{E}_0 = \begin{pmatrix} A_x \\ A_y e^{i\delta} \end{pmatrix}, \tag{6}$$

where A_x and A_y are the amplitude coefficients and δ is the phase difference between the two components [67,68]. Since conventional detectors only respond to the intensity of the incident light (i.e., $I \propto A_x^2 + A_y^2$), the phase difference δ, an important parameter, is inherently difficult to probe experimentally. To remedy this shortcoming of losing the phase information, the Stokes parameters are introduced to fully describe both the amplitude and SOP of a plane wave. Typically, the Stokes parameters are determined by six intensity measurements, which can be written as:

$$s_0 = A_x^2 + A_y^2 \tag{7}$$

$$s_1 = A_x^2 - A_y^2 \tag{8}$$

$$s_2 = 2A_x A_y \cos \delta = A_a^2 - A_b^2 \tag{9}$$

$$s_3 = 2A_x A_y \sin \delta = A_r^2 - A_l^2, \tag{10}$$

where s_0 is the intensity of the analyzed beam, and s_1–s_3 contain information on the SOP [67,68]. Additionally, s_1–s_3 can be obtained by measuring the intensities of the two orthogonal components in the three basis sets $(\hat{x}, \hat{y})$, $(\hat{a}, \hat{b}) = \frac{1}{\sqrt{2}}(\hat{x} + \hat{y}, -\hat{x} + \hat{y})$, and $(\hat{r}, \hat{l}) = \frac{1}{\sqrt{2}}(\hat{x} + i\hat{y}, \hat{x} - i\hat{y})$. Here, the basis $(\hat{a}, \hat{b})$ corresponds to a rotation of the Cartesian coordinate system $(\hat{x}, \hat{y})$ by 45° with respect to the x-axis, while $(\hat{r}, \hat{l})$ is the basis for CP light. It should be noted that s_1–s_3 are often normalized by s_0 so that all possible values are within ± 1. Additionaly, one can see that $(s_1^2 + s_2^2 + s_3^2)/s_0^2 = 1$, which shows that the SOPs in three-dimensional (3D) space (s_1, s_2, s_3) can be represented by the so-called Poincaré sphere.

3. Metasurface-Based Polarimeters

Metasurfaces represent an opportunity for polarimetry to overcome bulky and the expensive architectures of conventional volume optics. Here, we try to give an overview of recent progress in metasurface-based polarimeters during the past few years.

3.1. PB-Phase Metasurfaces for Determining Part of the SOP

In early approaches, PB-phase metasurfaces together with conventional optical elements, such as polarizers and retardation waveplates, were designed to determine the degree of circular polarization [69–71]. One example is depicted in Figure 2a, where an ultrathin (40 nm) gradient metasurface was formed and demonstrated to measure the ellipticity and handedness of polarized light [70]. Figure 2a shows the schematic and scanning electron microscope (SEM) image of the designed PB-phase metasurface, which consists of Au nanorods—with identical geometrical parameters but with spatially varying orientation—on top of anindium tin oxide (ITO) coated glass substrate. The angular orientation of each nanorod varies along the x-direction with an incremental $\pi/8$ clockwise rotation, but remains invariant in the y-direction. Hence, each period in x-direction contains eight nanorods, resulting in a phase shift ranging from 0 to 2π. Due to the spin-selected opposite slope of the PB-phase gradient, the decomposed RCP and LCP beams are steered in opposite directions, exhibiting the photonic spin Hall effect (PSHE) [72]. Thus, the obtained intensity distribution of the anomalously refracted LCP and RCP light provides an accurate and simple method to measure the ellipticity of the incident light. To experimentally demonstrate the capability of determining the degree of circular polarization with the proposed metasurface approach, various polarization states of incident light were impinged on the metasurface by changing the angle β between the axis of polarization and the fast axis of the quarter waveplate. These show different intensity distributions after interacting with the metasurface, as shown in Figure 2b. Figure 2c clearly shows that the ellipticity and handedness of various incident polarization states can be well characterized, and are in good agreement with the predicted values. Additionally, the proposed method here is suitable to work at a wide range of wavelengths, ascribed to the broadband nature of PB-phase. It should be noted that the conversion efficiency of the metasurface is strictly limited and the maximum conversion efficiency is around 7.6% at 940 nm, which may affect the performance of the proposed metasurface since the ellipticity and handedness of various incident polarization states are determined by the intensities of the refracted beams. The limited efficiency is ascribed to the single electric dipole resonance supported by the non-magnetic Au nanorods on top of ITO-coated glass substrate, which sets the upper bound of the coupling efficiency between the two polarizations to be 25% in the limit of negligible absorption [57,58].

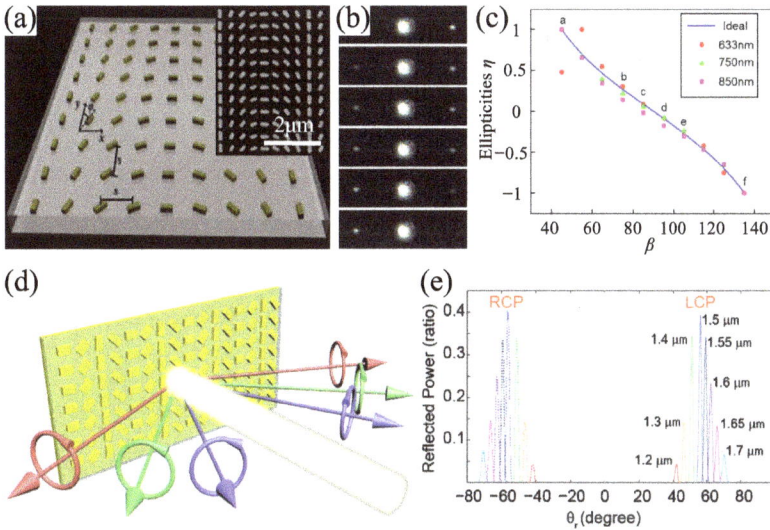

Figure 2. Pancharatnam–Berry (PB) phase metasurfaces for determining the degree of circular polarization. (**a**) Schematic illustration of the phase gradient metasurface. Inset shows the SEM image of the fabricated metasurface on an ITO-coated glass substrate; (**b**) Charge-coupled device (CCD) images for wavelength $\lambda_0 = 750$ nm and different angles β. (**c**) Experimentally obtained ellipticity η versus the incident polarization as a function of β; (**a–c**) Reproduced with permission from [70], Copyright Optical Society of America, 2015; (**d**) Illustration of the metasurface used as a circular dichroism spectrometer using the the photonic spin Hall effect (PSHE); (**e**) Experimental results of reflected power for left circularly polarized (LCP) and right circularly polarized (RCP) incident beams at different wavelengths as a function of reflected angle, showing the discrimination of LCP and RCP spectra; (**d,e**) Reproduced with permission from [71], Copyright Optical Society of America, 2015.

PB-phase GSPM have been demonstrated to increase the efficiency [71]. Figure 2d presents a novel and compact circular dichroism (CD) spectrometer based on PB-phase GSPM, which utilizes PSHE in reflection. Specifically, an array of anisotropic elements is used to achieve different phase gradients in response to LCP and RCP light. Therefore, the spin components of the incident broadband source are reflected in opposite directions, and the reflection angle of each wavelength component is varied according to generalized Snell's law [6]. When a broadband light source is incident on the metasurface, the LCP and RCP spectra are spatially separated and can be recorded simultaneously, which eliminates the need of switching the incident polarization, as shown in Figure 2e. By analyzing the spectra of different spin components, the CD of the light source can be easily obtained, which can serve as an important tool in sensing chiral molecules. Additionally, there is no other hardware required for this type of CD spectrometry.

As a final comment, it is worth pointing out that the obtained SOP of the incident light is incomplete as this type of PB-phase metasurface-based chiroptical spectroscopy has only two channels and cannot probe the polarization azimuthal angle. To fully characterize the SOP and determine the Stokes parameters, an additional polarizer is needed, which will inevitably increase the complexity of the whole system.

3.2. Metasurface-Only Polarimeters

In recent years, on-chip metasurface-only polarimeters have been proposed and were demonstrated to determine the entire SOP simultaneously. These polarimeters involve different configurations, including gap surface plasmon resonators [67,73], metallic nanoantennas [74,75], and all-dielectric nanoposts [76].

Starting with GSPMs, Figure 3a conveniently illustrates the basic working principle of the metagrating-based polarimeter, composed of three interweaved metasurfaces. An arbitrary polarized incident beam is spatially diffracted into six different directions, corresponding to different polarization states, since each metasurface functions as a polarization splitter for a certain polarization basis $(|x\rangle,|y\rangle)$, $(|a\rangle,|b\rangle)$, or $(|r\rangle,|l\rangle)$ [67]. As such, the proposed metagrating responds uniquely to all possible polarization states and the most pronounced differences in the corresponding diffractions occur for the six extreme polarizations $|x\rangle$, $|y\rangle$, $|a\rangle$, $|b\rangle$, $|r\rangle$, and $|l\rangle$. Hence, the entire SOP can be determined. By conducting simultaneous measurements of the corresponding far-field diffraction intensities in the six predesigned directions, the Stokes parameters can be quickly retrieved, thereby allowing one to easily analyze the SOP of the incident light. At the design wavelength of 800 nm, the experimental diffraction contrasts obtained by averaging three successive measurements are in good agreement with the input Stokes parameters, and lay on the Poincaré sphere, with points covering all octants of the 3D parameter space (Figure 3b). The two-norm deviation between Stokes parameters and experimental diffraction contrasts is estimated to be around ~0.1, which can be significantly reduced with better fabrication facilities. Importantly, due to the rather broadband response of the GSPM, the designed metagrating could work over a wide wavelength range. In the wavelength range of 750–850 nm, there is no significant degradation of the metagrating performance observed in the experiments; the experimentally diffraction contrasts closely represent the Stokes parameters.

In a later work [74], an ultracompact in-line polarimeter was demonstrated for transmission-type polarimetry by using a two-dimensional (2D) metasurface made up of a thin array of subwavelength metallic antennae embedded in a polymer film (Figure 3c). By placing two pairs of gold nanorod rows superimposed at a 45° relative angle, incident beams with four elliptical polarization states are directionally scattered. Based on the measured polarization-selective directional scattering in four directions from the corresponding out-coupling gratings, the SOP could be precisely obtained after calibration. From Figure 3d, one can clearly see that the measurement of several arbitrarily selected polarizations using the metasurface polarimeter agrees well with the result from the commercial polarimeter at a wavelength of 1550 nm. Considering the potential compactness, speed, and stability, this design is clearly superior to the commercial polarimeter. Following this concept, a compact fiber-coupled polarimeter with high sampling rates wasdemonstrated [75]. Here, we note that the degree of polarization (DOP) cannot be measured in the present four-output design. For DOP measurements, a more complex antenna array design is needed [74].

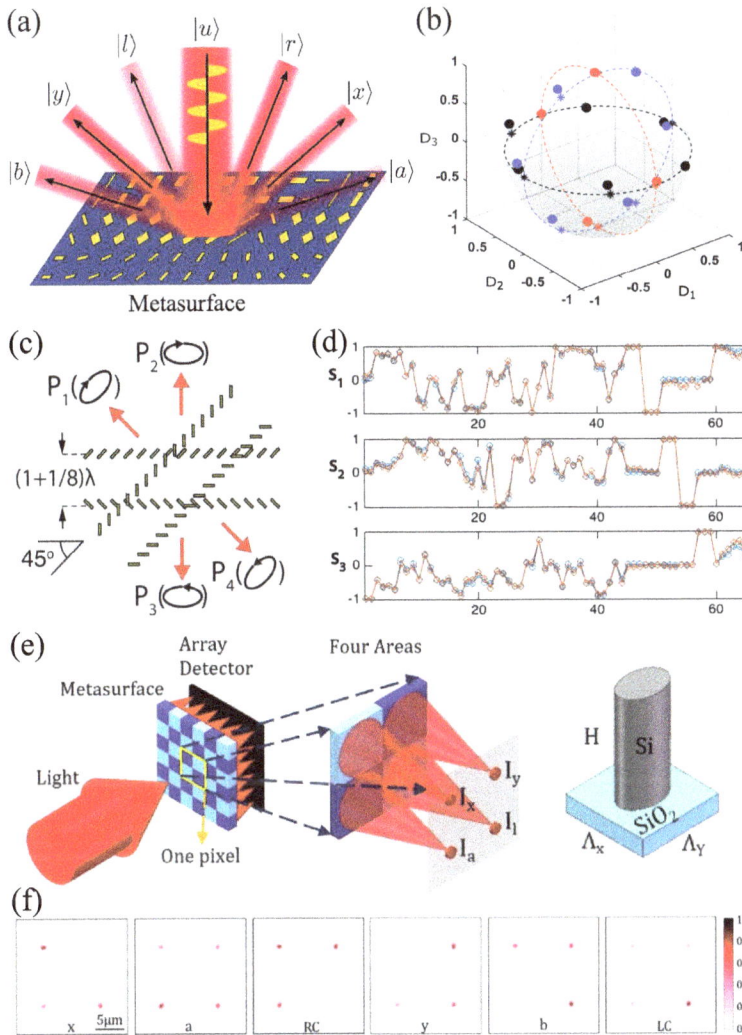

Figure 3. Metasurface-only polarimeters. (**a**) Illustration of the metagrating's working principle; (**b**) Measured diffraction contrasts (denoted by filled circles) for polarization states along the main axes of the Poincaré sphere (indicated by asterisks) at 800 nm; (**a,b**) Reproduced with permission from [67], Copyright Optical Society of America, 2015; (**c**) Polarization-selective directional scattering of four elliptical polarization states by two pairs of rows superimposed at a 45° relative angle; (**d**) Measurement of the state of polarization of arbitrarily selected polarizations using the commercial polarimeter (blue) and the metasurface polarimeter (orange) at 1550 nm; (**c,d**) Reproduced with permission from [74], Copyright Optical Society of America, 2016; (**e**) Schematic of ultracompact polarimeters based on dielectric metasurfaces. The inset shows the unit cell; (**f**) Simulated intensity distributions in the focal plane for *x*, *y*, *a*, *b*, RCP, and LCP incident waves; (**e,f**) Reproduced with permission from [76], Copyright Optical Society of America, 2017.

As complementary to compact polarimeters with plasmonic metasurfaces, all-dielectric metasurfaces may be tailored to design ultracompact polarimeters [76]. Compared with the GSPM-based polarimeters, such all-dielectric polarimeters could relax the difficulty of system integration as it is operating in transmission-mode, while maintaining high efficiency due to the relatively high refractive index and negligible absorption [22]. One example of a dielectric metasurface integrated polarimeter is depicted in Figure 3e, which contains a silicon metasurface and array detector with a distance of 4.2 μm corresponding to the focal plane of the metasurface. The metasurface consists of multiple pixels, where each pixel has four different areas containing periodically arranged elliptical silicon pillars resting on a silica substrate. Within the pixel, each area functions as a polarization-sensitive focusing lens, separating four polarization components from the incident light and focusing them to the detector's surface. Upon the excitation of a normally incident beam with arbitrary SOP, four focal spots are generated in the focal plane, which show different intensity distributions for different polarizations, as shown in Figure 3f. By detecting the intensities of the focused four polarization components in the focal plane (I_x, I_y, I_a, and I_l, which refer to the intensities of horizontal (H), vertical (V), +45°, and LCP components, respectively) and carefully calibrating the system, the Stokes parameters can be retrieved to fully describe the SOP of the incident light. Since each part of the incident light is fully used to determine the SOP, the detection efficiency only depends on the focusing efficiency of each flat focusing lens, which is theoretically estimated to be above 60%.

3.3. Metacoupler-Based Polarimeter

Inspired by the polarization-controlled unidirectional excitation of surface plasmon polaritons (SPPs) [33], Pors et al. suggested a metasurface-based polarimeter that can effectively couple normally incident light to in-plane plasmonic waveguide modes. By calculating the relative efficiency of excitation between predefined propagation directions, the incident SOP can be directly retrieved. Similar to the work in Reference [68], three properly designed GSPMs made up of metal-insulator-metal (MIM) nanoantennae embedded in a polymer layer were incorporated to make the so-called waveguide metacoupler, as depicted in Figure 4a,b. This metacoupler unidirectionally excites the plasmonic waveguide modes instead of far-field reflection propagating in six different directions for the three polarization basis sets—($|x\rangle,|y\rangle$), ($|a\rangle,|b\rangle$), and ($|r\rangle,|l\rangle$)—dictated by the definition of the Stokes parameters. Figure 4c–e displays the intensity distributions in an area of $30 \times 30 \, \mu m^2$ for three extreme SOPs, namely, $|x\rangle$, $|a\rangle$ and $|r\rangle$ polarizations, thereby illustrating all-polarization sensitivity. It is obvious that the metacoupler can launch the waveguide modes propagating in the six designed directions. Additionally, the power distribution in the six channels are strongly polarization-dependent and one of the six channels is suppressed depending on the input SOP. For example, port #1 is greatly suppressed for $|x\rangle$ polarization (Figure 4c) while port #4 becomes weak when the polarization is switched to $|a\rangle$ (Figure 4d). By integrating the power flows of the waveguide modes at the ports marked in Figure 4c, all the coupling efficiencies can be calculated, through which the normalized contrast denoted D_1–D_3 is determined. Here it should be noted that, unlike the previous work [67], there is no direct mathematical equivalence between D_1–D_3 and s_1/s_0–s_3/s_0, thus the retrieved contrasts can only roughly represent the input Stokes parameters. In order to achieve the best performance of the polarimeter, one needs to properly relate the six coupling efficiencies to the three Stokes parameters with calibration. After calibration, the proposed in-plane polarimeter shows excellent performance, with the retrieved results perfectly overlapping with the Stokes parameters, as displayed in Figure 4f.

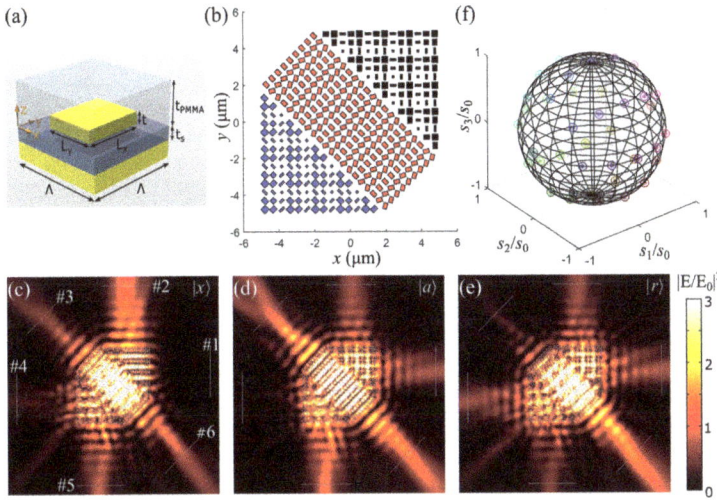

Figure 4. Metacoupler-based polarimeter. (**a**) Schematic of the unit cell; (**b**) Top view of the combined waveguide metacoupler; (**c–e**) Color map of the intensity in the center of the Poly(methyl methacrylate) (PMMA) layer for the metacoupler when the incident light is a Gaussian beam with a beam radius of 6 μm. The polarization state $|u\rangle$ of the beam is displayed in the upper right corner. Note that the scale bar is chosen to better highlight weak intensity features. Panel (**c**) shows the numbering of the six ports marked with gray lines; (**f**) Circles and asterisks indicate retrieved and exact polarization states of the incident beam for 42 different SOPs, plotted in (s_1, s_2, s_3)-space together with the Poincaré sphere. Reproduced with permission from [68], Copyright American Physical Society, 2016.

3.4. Metasurfaces Spectropolarimeters

In addition to determining the SOP of an incident beam, spectral analysis is concomitantly required. This resulted in the recent developments of metasurface-based spectropolarimeters, which enable simultaneous measurements of the spectrum and SOP [77–80]. As such, the spectropolarimeters are superior to the isolated spectrometers or polarimeters, regarding their capability of combining the uncorrelated information channels of intensity, wavelength, and SOP together.

Starting with segmented spectropolarimeters [77,78], a spectropolarimeter metadevice that steers different polarization and spectral components into predesigned spatial directions was proposed [77], which consists of six GSPMs arranged in a 2 × 3 array, corresponding to horizontal (0°), vertical (90°), ±45°, RCP, and LCP analyzers. Once a probe beam with a certain polarization state is incident on the metadevice, six diffraction spots are generated in the far-field, resulting from the anomalous reflection of each GSPM area. When the incident polarization is altered, the relative intensity distributions of diffraction spots change accordingly. As such, the polarization response of this spectropolarimeter at a given wavelength can be determined by carefully analyzing the relationship between the intensity of individual peaks and the incident SOP. Additionally, this metadevice could be used to analyze the wavelength due to the fact that the GSPMs are usually dispersive. The measured angular dispersion for the LCP and RCP channels are 0.053 °/nm and 0.024 °/nm, respectively. Hence, proving the potential of spectral measurement with spectral resolution up to ∼0.3 nm if this metadevice is inserted into a typical spectrometer setup. It should be noted that the spectral resolution of ∼0.3 nm is estimated with the aid of other optical components in a commercial spectrometer and the intrinsic resolution accessed with the metadevice itself is rather limited.

Although the metadevice with rectangular configuration shows good performance, it is relatively sensitive to the illuminated area by the incident light. In particular, the incident light should cover the whole area of the metasurface with the same light intensity in order to ensure faithful comparison of the corresponding diffraction orders. To increase the detection robustness and the compatibility with a circular laser beam that has a Gaussian profile, a segmented plasmonic spectropolarimeter has been demonstrated by using a center-symmetric configuration, featuring a self-calibrating nature [78]. The working principle of the self-calibrated spectropolarimeter is displayed in Figure 5a, which consists of three diffident types of GSPMs occupying 120° circular sectors each. Similar to the previous work [67], each sector-shaped GSPM operates as an efficient polarization splitter for one of the three polarization bases ($|x\rangle,|y\rangle$), ($|a\rangle,|b\rangle$), and ($|r\rangle,|l\rangle$). Upon the excitation of a normally incident monochromatic beam, this center-symmetrical configuration would diffract any normally incident beam to six predesigned directions, whose contrasts in the corresponding diffraction intensities would provide a direct measure of the SOP and retrieval of the associated Stokes parameters, due to the direct mathematical equivalence between the normalized contrasts and Stokes parameters. At the same time, the polar angle of diffraction is approximately proportional to the wavelength, resulting in the spectral analysis of the incident beam. The fabricated 96 μm diameter spectropolarimeters, operating in the wavelength range of 750–950 nm, are found to exhibit excellent capability of polarization detection (Figure 5b). Moreover, the experimentally measured angular dispersion $\Delta\theta/\Delta\lambda$ for the $|x\rangle$ channel is 0.0133 °/nm, corresponding to a measured spectral resolving power of $\lambda/\Delta\lambda \approx 15.2$ (Figure 5c,d). Importantly, due to the circular-sector design, polarization analysis can be conducted for optical beams of different diameters without prior calibration, thereby demonstrating the beam-size invariant functionality. As a final comment, we would like to stress that uneven illumination of the three metasurfaces caused by misalignment or inhomogeneity of the laser beam will not affect the retrieved Stokes parameters and no calibration is needed since the Stokes parameters are only related to the relative diffraction contrasts for three polarization bases.

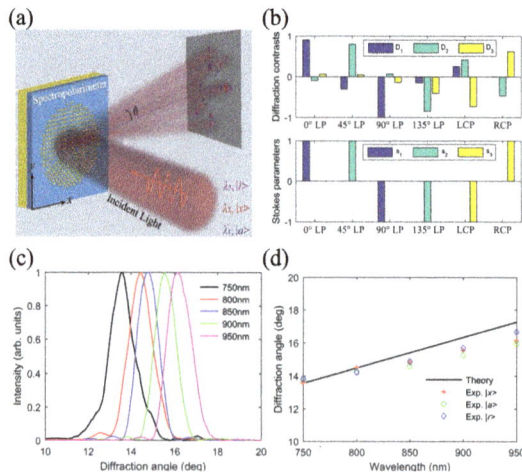

Figure 5. Segmented spectropolarimeters. (**a**) Illustration of the gap-surface plasmon metasurface (GSPM)-based beam-size invariant spectropolarimeter; (**b**) Measured diffraction contrasts for the six extreme polarizations that represent the values of Stokes parameters at $\lambda = 800$ nm; (**c**) Normalized measured far-field intensity profile for different wavelengths of the $|x\rangle$ channel; (**d**) Theoretical and measured spectral dispersions of different channels. Reproduced with permission from [78], Copyright American Chemical Society, 2017.

To increase the spectral resolution of spectropolarimeters, interleaved metasurfaces have been designed to conduct spectropolarimetry, where different channels share the aperture of the device [79,80]. A good example of interleaved metasurface-based spectropolarimeters is shown in Figure 6a, where three linear phase profiles associated with different nanoantenna subarrays, formed by the random interspersing, make up the configuration. This enables simultaneous characterization of the SOP and spectrum in the reflected beam [79]. In particular, by randomly mixing optical nanoantenna arrays with different phase functions, the PB-phase concept, spatially interleaved phase profiles are designed, which yield extraordinary information capability. When a probe beam with an arbitrary SOP is incident on the spectropolarimeter metasurface (SPM), two beams of intensities $I_{\sigma+}$ and $I_{\sigma-}$, consisting of opposite helicity states, and two additional beams emerge. The latter two beams are projected onto linear polarizers at 0° and 45°, determining the linearly polarized components I_{L0} and I_{L45}, respectively. The Stokes parameters of the incident beam are then calculated with a calibration experiment. Figure 6b shows the measured and calculated Stokes parameters on a Poincaré sphere for an analyzed beam impinging the SPM at a wavelength of 760 nm with different polarizations, demonstrating the excellent capability of the polarization probe. Furthermore, such interleaved SPM shows good spectral resolving power, which was measured to be $\lambda/\Delta\lambda \approx 13$ when the diameter is only 50 μm (Figure 6c). To further improve the spectral resolving power while keeping the aperture size fixed, super-dispersive, off-axis metalenses that could simultaneously disperse and focus light of different wavelengths can be used [81,82]. Based on this SPM, the optical rotatory dispersion (ORD) for the specific rotations of D-glucose (chiral molecule) and its enantiomer L-glucose were measured. The ORD of D-glucose shows good agreement with the values found in the literature and the L-glucose ORD is manifested by the opposite behavior as expected (Figure 6d). Finally, we note that this interleaved center-symmetrical approach can be applied with dielectric metasurfaces for spectropolarimetry operating in the visible wavelength range [80].

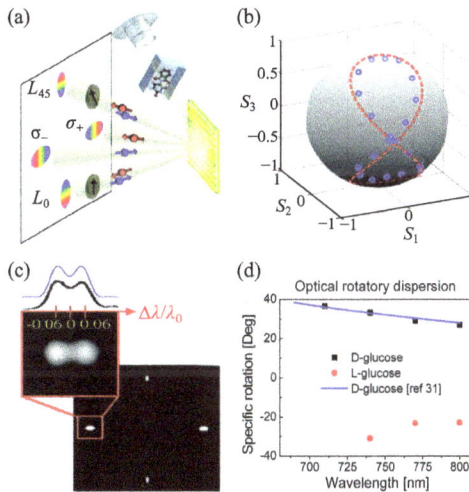

Figure 6. Interleaved spectropolarimeter. (**a**) Schematic setup of the spectropolarimeter. The spectropolarimeter metasurface (SPM) is illuminated by continuum light passing through a cuvette with chemical solvent, then four beams of intensities $I_{\sigma+}$, $I_{\sigma-}$, I_{L45}, and I_{L0} are reflected toward a CCD; (**b**) Predicted (red dashed curve) and measured (blue circles) polarization states, depicted on a Poincaré sphere; (**c**) Measured far-field intensities for elliptical polarization at two spectral lines (with wavelengths of 740 and 780 nm) and (inset) the corresponding resolving power (black line) and calculation (blue line) of the 50 μm diameter SPM; (**d**) Optical rotatory dispersion (ORD) for the specific rotations of D- and L-glucose. Black squares and red circles represent the measured ORD of D- and L-glucose, respectively. The blue line depicts the dispersion acquired from the literature. Reproduced with permission from [79], Copyright American Association for the Advancement of Science, 2016.

3.5. On-Chip Spectropolarimetry by Fingerprinting with Random Metasurfaces

Metasurfaces composed of disordered structure have been demonstrated to realize various functions, such as diffusing light [29], absorbing light [83], waveguiding [84], wavefront shaping [85–87], and polarimetry [88]. In Reference [88], a unique on-chip spectropolarimeter, by fingerprinting with a random gold nanoparticle (NP) array, in the near-infrared range is presented. This spectropolarimeter is based on analyzing scattering SPP patterns from micron-sized circular arrays of randomly distributed gold NPs (Figure 7a) through a leakage radiation microscope (LRM). Due to strong multiple scattering and the countless scattering routes between the closely packed NPs, the disordered media is able to generate unique and complicated scattering patterns that are sensitive to the wavelength and SOP, and thereby those scattering spectra can be employed as spectropolarimetric fingerprints to distinguish incident light carrying different polarization and spectrum components. The random metasurface consists of gold NPs with diameter of about 50 nm, thickness of 70 nm, and density of 75 μm^{-2}, distributed in circular areas with radii of 1–5 μm on top of a 70 nm thick gold film and 170 μm thick silica substrate (Figure 7b). Larger array sizes generate more complex scattering patterns.

In the experiment, the angular spectra of the scattering patterns are characterized along the azimuthal angle θ through their Fourier plane images from the LRM (Figure 7c). The scattering spectra of different SOPs and wavelengths are quantitatively compared by calculating the correlation coefficients. The experimental results shows a substantial difference between the main SOPs: the orthogonal SOPs (horizontal and vertical linear, or left and right circular) exhibit virtually no correlation (<0.2). They also investigated the influence of the angle (α) between two different linear polarizations, and the correlation coefficient exhibits similar angular dependence as the analytical expression of the polarizer transmission ($\cos^2\alpha$) with the increasing polarization angle difference (Figure 7d). In the case of varying wavelengths, the correlation coefficient also drops accordingly with increasing wavelength difference in the 700–1000 nm range, with the wavelength selectivity estimated to be λ^2/L_{spp} (L_{spp} is the propagation length of the SPP) (Figure 7e). Numerical simulations of angular SPP scattering spectra are also carried out and confirm the sensitivity of the SPP scattering spectra to different wavelength and SOPs, thereby validating the proposed concept of on-chip spectropolarimetry by fingerprinting based on SPP excitation and multiple scattering by random surface nanostructures. Compared to other metasurface-based polarimetry, the fingerprinting approach might require extra work for the careful calibration; the registration of different SPP scattering spectra to create a database of spectropolarimetric fingerprints. However, it also has the advantage of simple design and fabrication procedure, in addition to its potential applications for optical wavefront sensing, optical beam tracing and positioning, and so on.

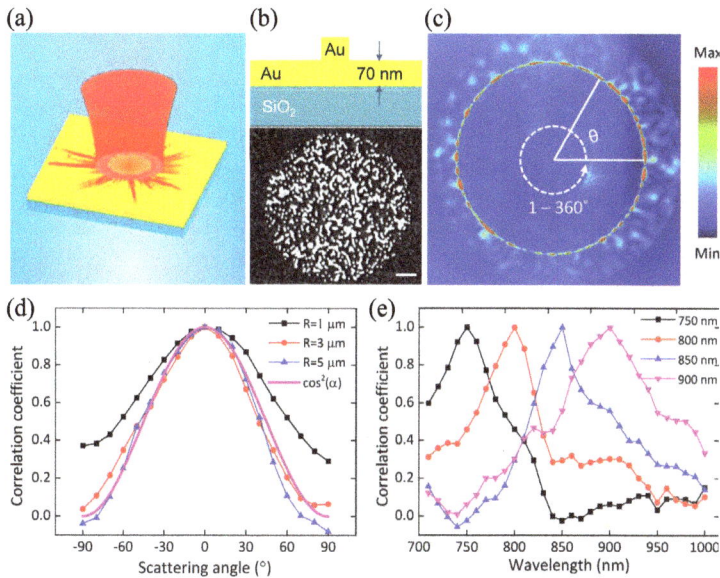

Figure 7. On-chip spectropolarimetry by fingerprinting with random metasurfaces. (**a**) Schematic of surface plasmon polariton (SPP) excitation and scattering occurring upon illumination of random nanoparticle (NP) arrays at normal incidence; (**b**) Sketch of an individual gold NP atop a 70-nm-thick gold film deposited on a silica wafer; (**c**) Corresponding leakage radiation microscopy (LRM) image in the Fourier plane introducing the azimuthal angle θ that denotes the SPP scattering direction. The bright circle is formed by SPP waves scattered in different directions with its radius being related to the SPP effective index; (**d**) Correlation between SPP scattering spectra obtained for different linear polarizations (at 800 nm) with differently sized random NP arrays as a function of the angle between the polarizations; (**e**) Correlation between SPP scattering spectra obtained with linear horizontal polarization as a function of the incident wavelength, when using different reference wavelengths of 750, 800, 850, and 900 nm. Reproduced with permission from [88], Copyright American Chemical Society, 2018.

3.6. Metasurface OAM Spectropolarimeters

Besides the spectrum and polarization, the orbital angular momentum (OAM) is also an intrinsic property of light, which processes a helical phase front, such that the Poynting vector within the beam is twisted with respect to the principal axis of light propagation [89]. In contrast to the spin angular momentum (SAM) that can take only two values, OAM is unbounded since the topological charge l can take any integer value. Thus OAM beams have recently gained tremendous interest in optical trapping, high-resolution microscopy, and quantum information processing, as they introduces an additional degree of freedom for encoding the light beam and increasing communication capabilities [20,90]. A variety of metasurfaces have been demonstrated to generate OAM beams [20,91–94], which have, in turn, motivated the implementation of OAM polarimetry, enabling direct detection of topological changes to a structured optical wavefront. Very recently, Hasman's group incorporated spectropolarimetry and OAM sensing into a silicon metasurface, which would ultimately facilitate a simultaneous detection of the wavelength, polarization, and OAM of light by projecting an incident mode $|l_{in}\rangle$ on a set of orthogonal OAM states $|l_j\rangle$ [80].

The OAM spectropolarimeter metasurface (OSPM), designed with the combination of the shared-aperture and PB-phase concepts, consists of interleaved antenna sub-arrays with multiplexed geometric-phase profiles. These result in a finite number of dominant multiplexed OAM harmonic orders with identical intensities. Illuminating the OSPM with an arbitrarily polarized polychromatic light source results in four symmetric sets of annular spots with a bright spot at the center, as shown in Figure 8a. Given an incident beam possessing an arbitrary topological charge l_{in}, the winding numbers of the diffracted orbital harmonics are modified by accumulating the value of l_{in}. This shifts the bright spot from its original location $l = 0$ to $l = -l_{in}$, thus enabling the determination of the incident OAM value (Figure 8b–g). From the far-field images, one can clearly see that the position of the bright spot changes in response to alternating the incident OAM. For instance, the bright spot moves to the right side if the incident OAM carries a topological change of $l_{in} = 1$ (Figure 8c), while the bright spot moves to the left side if $l_{in} = -1$ (Figure 8e). Additionally, such OSPM can resolve the incident polarization states and wavelengths precisely.

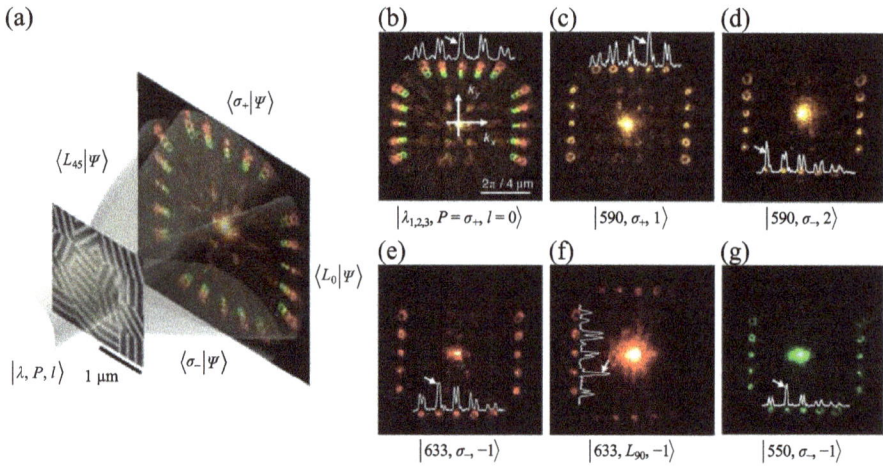

Figure 8. Silicon-based orbital angular momentum (OAM) Spectropolarimeter. (**a**) Schematic set-up of an OAM spectropolarimeter metasurface (OSPM) of 50 µm diameter illuminated by an arbitrarily polarized polychromatic light source. The image captured by the CCD was obtained for an elliptically polarized beam at $\lambda_{1,2,3}$ = 550, 590 and 633 nm with topological charge, $l = 0$; (**b–g**) Measured far-field intensities for different wavelengths, polarizations, and OAMs. Note, $|\lambda, P, l\rangle$ stands for the incident beam state, representing the wavelength, the polarization, and the OAM, respectively. Reproduced with permission from [80], Copyright Macmillan Publishers Ltd: Nature Light: Science and Applications, 2017.

3.7. Photodetector Integrated Polarimeters

In the previous sections, the SOPs are mainly determined by the measured optical signals. An as-yet unexplored milestone in the field of metasurface-based polarimeters is to achieve an integrated electronic device, which can electronically probe the polarization states. Owing to the fascinating properties of metasurfaces, several photodetector integrated polarimeters have been proposed and demonstrated by integrating meta-atoms with semiconductor elements [95,96]. In this section, we will review some of the recent progress in photodetector integrated polarimeters.

As a good example, Figure 9a schematically shows a silicon photodetector integrated with a set of plasmonic structures that can be used as either a broadband linear-Stokes polarimeter or a narrowband full-Stokes polarimeter, capable of determining the complete state of polarization of a light beam [95]. Specifically, in this silicon-based Schottky detector, the Schottky barrier is created by using a few nanometer thick chromium layer between the n-type silicon substrate and the gold layer. The gold contact is patterned with four linear subwavelength slits at different orientations and two subwavelength coaxial apertures surrounded by spiral grooves with opposite twists, which have different optical responses to the incident polarizations. In this way, these differently shaped plasmonic structures patterned in the Au film can work as the polarization filters for the six extreme polarizations ($|x\rangle$, $|y\rangle$, $|a\rangle$, $|b\rangle$, $|r\rangle$ and $|l\rangle$). Given the linear relationship between the generated photocurrent and the light intensity behind the filters in the silicon substrate, the required intensity measurements can effectively be translated into a set of photocurrent measurements. Figure 9b shows the fabricated linear Stokes detector in which four linear slits are milled into the gold film with focused ion beam (FIB) milling. By raster scanning the laser beam over the device, the photocurrent images of the linear Stokes detector at $\lambda = 830$ nm are obtained, confirming that the light absorption in the silicon is dominated by the portion of the light polarized normal to each slit (Figure 9c). The photocurrent response of a single slit at a given polarization is defined as the maximum measured photocurrent found in scanning the beam over the slit. If the direction of the incident electric field is continuously varied with a half wave plate, the photocurrent response of a single slit changes accordingly, revealing the linear polarization-dependent response of the slit, as shown in Figure 9d. Therefore the arrangement of four slits can be used for linear Stokes polarimetry with which one can distinguish between linear, circular, and elliptical states of polarization and determine the degree of linear polarization for partially polarized light over a broadband wavelength range. To determine the handedness of a circularly or elliptically polarized light and also degree of circular polarization for partially polarized light, two subwavelength coaxial apertures surrounded by spiral grooves with opposite twists have been designed, which respond differently to CP light of opposite handedness due to geometric phase effects (Figure 9e,f) [97,98].

Very recently, a CP light detector was demonstrated by combining a chiral metasurface with hot electron injection, enabling a potential application for polarimetry [96]. As shown in Figure 9g, the proposed chiral metasurface—consisting of the chiral plasmonic meta-atom array, dielectric spacer, and metal backplane—can perfectly absorb CP light with one particular handedness while largely reflecting the opposite component. Within the metasurface, the continuous 'Z'-shaped silver antenna allows for an electrical connection with a silver bus serving as the electrode (Figure 9h). The whole device is realized by placing an n-type silicon wafer in contact with the antenna layer, forming a Schottky barrier, as shown in Figure 9i. As such, the device can selectively generate hot electrons and produce a photocurrent signal depending on the handedness of light. By illuminating the metasurfaces with a CP laser and measuring photocurrent as a function of the laser handedness and wavelength, the photoresponse spectra of the devices were obtained (Figure 9j,k). In general, the photoresponsivity spectrum matches well with the measured absorption spectrum. When the metasurface is left-handed, it absorbs LCP light and reflects RCP light, resulting in dominating photoresponsivity for LCP light. Once the handedness is switched to right, it generates a stronger current signal for RCP light. Furthermore, the large distinction in the photocurrent for LCP and RCP light corresponds to the large CD (Figure 9l). Therefore, by including both right-handed and left-handed surface patterns, the sensor can differentiate between LCP and RCP light.

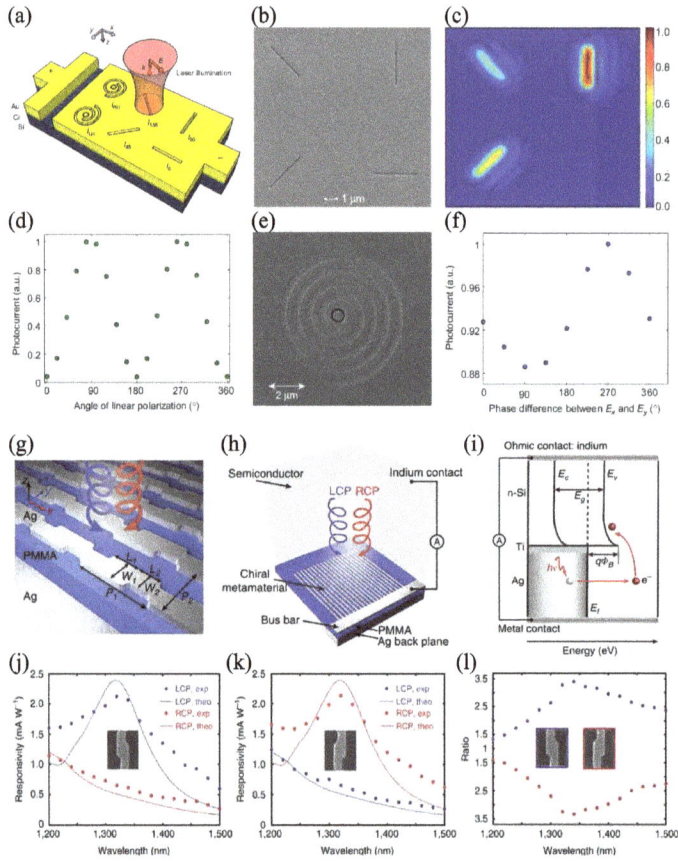

Figure 9. Photodetector integrated polarimeters. (**a**) Schematic of the proposed plasmonic polarimeter that consists of six differently shaped plasmonic slit structures patterned into a gold film on top of a silicon-based Schottky detector; (**b**) SEM image of a part of the Stokes detector; (**c**) Measured photocurrent map of the linear Stokes detector for a linearly polarized incident beam. The angle between the incident electric field and the horizontal slit is 7° and the illumination wavelength is 830 nm. The photocurrent is normalized to the maximum measured photocurrent; (**d**) Normalized measured photocurrent of the horizontal slit as a function of the polarization angle of the incident linearly polarized light; (**e**) SEM image of a coaxial aperture surrounded by an right-handed spiral groove fabricated with focused ion beam (FIB) milling; (**f**) Normalized measured photocurrent of the coaxial aperture shown in (**e**) as a function of the phase difference between E_x and E_y; (**a**–**f**) Reproduced with permission from [95], Copyright Science Wise Publishing § De Gruyter Berlin, 2012; (**g**) Schematic of the chiral metasurface consisting of a chiral plasmonic meta-molecule array, dielectric spacer, and metal backplane; (**h**) Schematic of the circularly polarized (CP) light detector consisting of a chiral metasurface integrated with a semiconductor that serves as a hot electron acceptor; (**i**) Energy band diagram of the CP light detector; (**j,k**) Experimentally measured (dots) and theoretically calculated (solid curve) photoresponsivity spectra under LCP (blue) and RCP (red) illumination for left-handed (**j**) and right-handed (**k**) metasurfaces; (**l**) Photocurrent polarization discrimination ratio spectra of left-handed and right-handed metasurfaces; (**g**–**l**) Reproduced with permission from [96], Copyright Macmillan Publishers Ltd.: Nature Communications, 2015.

4. Conclusions and Outlook

We have overviewed the rapid development of metasurface-based polarimeters. From the perspectives of physical mechanisms, the concepts of generalized Snell's law, PB-phase, and Stokes parameters were introduced. After that, we reviewed the efforts of metasurface-based polarimeters, including generic polarimeters, spectropolarimeters, OAM spectropolarimeters, and photodetector integrated polarimeters.

Although much progresses has been achieved in the field of metasurface-based polarimeters, there are still some challenges to overcome in future. To date, most of metasurface-based polarimeters destroy or substantially modify the original wavefronts of an incident beam in the detection process. Therefore, a nondestructive and real-time polarimeter is desired, which leaves the original wavefronts virtually unaffected [99]. Additionally, nanofabrication of the metasurface-based polarimeters mainly relies on high-cost and time-consuming electron beam lithography or FIB milling, which may hinder the potential applications of metasurface optical devices. Nanoimprinting should be a feasible way to the low cost and large area fabrication of metasurfaces [100].

Acknowledgments: This work was funded by the European Research Council (the PLAQNAP project, Grant 341054) and the University of Southern Denmark (SDU2020 funding).

Author Contributions: Fei Ding and Yiting Chen contributed to writing and finalizing the paper. Sergey I. Bozhevolnyi supervised the project.

Conflicts of Interest: The authors declare no conflict of interest.

References

1.   Born, M.; Wolf, E. *Principles of Optics: Electromagnetic Theory of Propagation, Interference and Diffraction of Light*; Elsevier: New York, NY, USA, 2013.
2.   Sterzik, M.F.; Bagnulo, S.; Palle, E. Biosignatures as revealed by spectropolarimetry of Earthshine. *Nature* **2012**, *483*, 64–66.
3.   Gupta, N. Acousto-optic-tunable-filter-based spectropolarimetric imagers for medical diagnostic applications—Instrument design point of view. *J. Biomed. Opt.* **2005**, *10*, 051802.
4.   Tyo, J.S.; Goldstein, D.L.; Chenault, D.B.; Shaw, J.A. Review of passive imaging polarimetry for remote sensing applications. *Appl. Opt.* **2006**, *45*, 5453–5469.
5.   Bohren, C.F.; Huffman, D.R. *Absorption and Scattering of Light by Small Particles*; Wiley-VCH Verlag GmbH: Weinheim, Germany, 2004.
6.   Yu, N.; Genevet, P.; Kats, M.A.; Aieta, F.; Tetienne, J.P.; Capasso, F.; Gaburro, Z. Light Propagation with Phase Discontinuities: Generalized Laws of Reflection and Refraction. *Science* **2011**, *334*, 333–337.
7.   Yu, N.; Capasso, F. Flat optics with designer metasurfaces. *Nat. Mater.* **2014**, *13*, 139–150.
8.   Meinzer, N.; Barnes, W.L.; Hooper, I.R. Plasmonic meta-atoms and metasurfaces. *Nat. Photon.* **2014**, *8*, 889–898.
9.   Koenderink, A.F.; Alù, A.; Polman, A. Nanophotonics: Shrinking light-based technology. *Science* **2015**, *348*, 516–521.
10.   Luo, X.; Pu, M.; Ma, X.; Li, X. Taming the Electromagnetic Boundaries via Metasurfaces: From Theory and Fabrication to Functional Devices. *Int. J. Antennas Propag.* **2015**, *2015*, 204127.
11.   Minovich, A.E.; Miroshnichenko, A.E.; Bykov, A.Y.; Murzina, T.V.; Neshev, D.N.; Kivshar, Y.S. Functional and nonlinear optical metasurfaces. *Laser Photon. Rev.* **2015**, *9*, 195–213.
12.   Shaltout, A.M.; Kildishev, A.V.; Shalaev, V.M. Evolution of photonic metasurfaces: From static to dynamic. *J. Opt. Soc. Am. B* **2016**, *33*, 501–510.
13.   Zhang, L.; Mei, S.; Huang, K.; Qiu, C.W. Advances in Full Control of Electromagnetic Waves with Metasurfaces. *Adv. Opt. Mater.* **2016**, *4*, 818–833.
14.   Glybovski, S.B.; Tretyakov, S.A.; Belov, P.A.; Kivshar, Y.S.; Simovski, C.R. Metasurfaces: From microwaves to visible. *Phys. Rep.* **2016**, *634*, 1–72.
15.   Chen, H.T.; Taylor, A.J.; Yu, N. A review of metasurfaces: Physics and applications. *Rep. Prog. Phys.* **2016**, *79*, 076401.

16. Hsiao, H.H.; Chu, C.H.; Tsai, D.P. Fundamentals and Applications of Metasurfaces. *Small Methods* **2017**, *1*, 1600064.

17. Ding, F.; Pors, A.; Bozhevolnyi, S.I. Gradient metasurfaces: A review of fundamentals and applications. *Rep. Prog. Phys.* **2018**, *81*, 026401.

18. Yu, N.; Genevet, P.; Aieta, F.; Kats, M.A.; Blanchard, R.; Aoust, G.; Tetienne, J.P.; Gaburro, Z.; Capasso, F. Flat Optics: Controlling Wavefronts With Optical Antenna Metasurfaces. *IEEE J. Sel. Top. Quantym Electron.* **2013**, *19*, 4700423.

19. Hum, S.V.; Perruisseau-Carrier, J. Reconfigurable Reflectarrays and Array Lenses for Dynamic Antenna Beam Control: A Review. *IEEE Trans. Antennas Propag.* **2014**, *62*, 183–198.

20. Guanghao, R.; Qiwen, Z. Tailoring optical complex fields with nano-metallic surfaces. *Nanophotonics* **2015**, *4*, 2–25.

21. Estakhri, N.M.; Alù, A. Recent progress in gradient metasurfaces. *J. Opt. Soc. Am. B* **2016**, *33*, A21–A30.

22. Jahani, S.; Jacob, Z. All-dielectric metamaterials. *Nat. Nanotechnol.* **2016**, *11*, 23–36.

23. Zheludev, N.I.; Kivshar, Y.S. From metamaterials to metadevices. *Nat. Mater.* **2012**, *11*, 917–924.

24. Genevet, P.; Capasso, F. Holographic optical metasurfaces: A review of current progress. *Rep. Prog. Phys.* **2015**, *78*, 024401.

25. Ni, X.; Emani, N.K.; Kildishev, A.V.; Boltasseva, A.; Shalaev, V.M. Broadband Light Bending with Plasmonic Nanoantennas. *Science* **2012**, *335*, 427.

26. Sun, S.; Yang, K.Y.; Wang, C.M.; Juan, T.K.; Chen, W.T.; Liao, C.Y.; He, Q.; Xiao, S.; Kung, W.T.; Guo, G.Y.; et al. High-Efficiency Broadband Anomalous Reflection by Gradient Meta-Surfaces. *Nano Lett.* **2012**, *12*, 6223–6229.

27. Pors, A.; Albrektsen, O.; Radko, I.P.; Bozhevolnyi, S.I. Gap plasmon-based metasurfaces for total control of reflected light. *Sci. Rep.* **2013**, *3*, 2155.

28. Pfeiffer, C.; Grbic, A. Metamaterial Huygens' Surfaces: Tailoring Wave Fronts with Reflectionless Sheets. *Phys. Rev. Lett.* **2013**, *110*, 197401.

29. Pors, A.; Ding, F.; Chen, Y.; Radko, I.P.; Bozhevolnyi, S.I. Random-phase metasurfaces at optical wavelengths. *Sci. Rep.* **2016**, *6*, 28448.

30. Sun, S.; He, Q.; Xiao, S.; Xu, Q.; Li, X.; Zhou, L. Gradient-index meta-surfaces as a bridge linking propagating waves and surface waves. *Nat. Mater.* **2012**, *11*, 426–431.

31. Lin, J.; Mueller, J.B.; Wang, Q.; Yuan, G.; Antoniou, N.; Yuan, X.C.; Capasso, F. Polarization-controlled tunable directional coupling of surface plasmon polaritons. *Science* **2013**, *340*, 331–334.

32. Huang, L.; Chen, X.; Bai, B.; Tan, Q.; Jin, G.; Zentgraf, T.; Zhang, S. Helicity dependent directional surface plasmon polariton excitation using a metasurface with interfacial phase discontinuity. *Light Sci. Appl.* **2013**, *2*, e70.

33. Pors, A.; Nielsen, M.G.; Bernardin, T.; Weeber, J.C.; Bozhevolnyi, S.I. Efficient unidirectional polarization-controlled excitation of surface plasmon polaritons. *Light Sci. Appl.* **2014**, *3*, e197.

34. Sun, W.; He, Q.; Sun, S.; Zhou, L. High-efficiency surface plasmon meta-couplers: Concept and microwave-regime realizations. *Light Sci. Appl.* **2016**, *5*, e16003.

35. Ding, F.; Deshpande, R.; Bozhevolnyi, S.I. Bifunctional gap-plasmon metasurfaces for visible light: Polarization-controlled unidirectional surface plasmon excitation and beam steering at normal incidence. *Light Sci. Appl.* **2018**, *7*, e17178.

36. Li, X.; Xiao, S.; Cai, B.; He, Q.; Cui, T.J.; Zhou, L. Flat metasurfaces to focus electromagnetic waves in reflection geometry. *Opt. Lett.* **2012**, *37*, 4940–4942.

37. Aieta, F.; Genevet, P.; Kats, M.A.; Yu, N.; Blanchard, R.; Gaburro, Z.; Capasso, F. Aberration-Free Ultrathin Flat Lenses and Axicons at Telecom Wavelengths Based on Plasmonic Metasurfaces. *Nano Lett.* **2012**, *12*, 4932–4936.

38. Ni, X.; Ishii, S.; Kildishev, A.V.; Shalaev, V.M. Ultra-thin, planar, Babinet-inverted plasmonic metalenses. *Light Sci. Appl.* **2013**, *2*, e72.

39. Pors, A.; Nielsen, M.G.; Eriksen, R.L.; Bozhevolnyi, S.I. Broadband Focusing Flat Mirrors Based on Plasmonic Gradient Metasurfaces. *Nano Lett.* **2013**, *13*, 829–834.

40. Arbabi, A.; Horie, Y.; Bagheri, M.; Faraon, A. Dielectric metasurfaces for complete control of phase and polarization with subwavelength spatial resolution and high transmission. *Nat. Nanotechnol.* **2015**, *10*, 937–943.

41. Khorasaninejad, M.; Chen, W.T.; Devlin, R.C.; Oh, J.; Zhu, A.Y.; Capasso, F. Metalenses at visible wavelengths: Diffraction-limited focusing and subwavelength resolution imaging. *Science* **2016**, *352*, 1190–1194.

42. Arbabi, A.; Arbabi, E.; Kamali, S.M.; Horie, Y.; Han, S.; Faraon, A. Miniature optical planar camera based on a wide-angle metasurface doublet corrected for monochromatic aberrations. *Nat. Commun.* **2016**, *7*, 13682.

43. Wang, S.; Wu, P.C.; Su, V.C.; Lai, Y.C.; Chu, C.H.; Chen, J.W.; Lu, S.H.; Chen, J.; Xu, B.; Kuan, C.H.; et al. Broadband achromatic optical metasurface devices. *Nat. Commun.* **2017**, *8*, 187.

44. Ni, X.; Kildishev, A.V.; Shalaev, V.M. Metasurface holograms for visible light. *Nat. Commun.* **2013**, *4*, 2807.

45. Chen, W.T.; Yang, K.Y.; Wang, C.M.; Huang, Y.W.; Sun, G.; Chiang, I.D.; Liao, C.Y.; Hsu, W.L.; Lin, H.T.; Sun, S.; et al. High-efficiency broadband meta-hologram with polarization-controlled dual images. *Nano Lett.* **2013**, *14*, 225–230.

46. Huang, L.; Chen, X.; Mühlenbernd, H.; Zhang, H.; Chen, S.; Bai, B.; Tan, Q.; Jin, G.; Cheah, K.W.; Qiu, C.W.; et al. Three-dimensional optical holography using a plasmonic metasurface. *Nat. Commun.* **2013**, *4*, 2808.

47. Zheng, G.; Mühlenbernd, H.; Kenney, M.; Li, G.; Zentgraf, T.; Zhang, S. Metasurface holograms reaching 80% efficiency. *Nat. Nanotechnol.* **2015**, *10*, 308–312.

48. Wen, D.; Yue, F.; Li, G.; Zheng, G.; Chan, K.; Chen, S.; Chen, M.; Li, K.F.; Wong, P.W.H.; Cheah, K.W.; et al. Helicity multiplexed broadband metasurface holograms. *Nat. Commun.* **2015**, *6*, 8241.

49. Khorasaninejad, M.; Ambrosio, A.; Kanhaiya, P.; Capasso, F. Broadband and chiral binary dielectric meta-holograms. *Sci. Adv.* **2016**, *2*, e1501258.

50. Yu, N.; Aieta, F.; Genevet, P.; Kats, M.A.; Gaburro, Z.; Capasso, F. A Broadband, Background-Free Quarter-Wave Plate Based on Plasmonic Metasurfaces. *Nano Lett.* **2012**, *12*, 6328–6333.

51. Pors, A.; Nielsen, M.G.; Bozhevolnyi, S.I. Broadband plasmonic half-wave plates in reflection. *Opt. Lett.* **2013**, *38*, 513–515.

52. Yang, Y.; Wang, W.; Moitra, P.; Kravchenko, I.I.; Briggs, D.P.; Valentine, J. Dielectric Meta-Reflectarray for Broadband Linear Polarization Conversion and Optical Vortex Generation. *Nano Lett.* **2014**, *14*, 1394–1399.

53. Ding, F.; Wang, Z.; He, S.; Shalaev, V.M.; Kildishev, A.V. Broadband High-Efficiency Half-Wave Plate: A Supercell-Based Plasmonic Metasurface Approach. *ACS Nano* **2015**, *9*, 4111–4119.

54. Wu, P.C.; Tsai, W.Y.; Chen, W.T.; Huang, Y.W.; Chen, T.Y.; Chen, J.W.; Liao, C.Y.; Chu, C.H.; Sun, G.; Tsai, D.P. Versatile polarization generation with an aluminum plasmonic metasurface. *Nano Lett.* **2017**, *17*, 445–452.

55. Pors, A.; Nielsen, M.G.; Valle, G.D.; Willatzen, M.; Albrektsen, O.; Bozhevolnyi, S.I. Plasmonic metamaterial wave retarders in reflection by orthogonally oriented detuned electrical dipoles. *Opt. Lett.* **2011**, *36*, 1626–1628.

56. Zhao, Y.; Alù, A. Manipulating light polarization with ultrathin plasmonic metasurfaces. *Phys. Rev. B* **2011**, *84*, 205428.

57. Monticone, F.; Estakhri, N.M.; Alù, A. Full control of nanoscale optical transmission with a composite metascreen. *Phys. Rev. Lett.* **2013**, *110*, 203903.

58. Ding, X.; Monticone, F.; Zhang, K.; Zhang, L.; Gao, D.; Burokur, S.N.; de Lustrac, A.; Wu, Q.; Qiu, C.W.; Alù, A. Ultrathin Pancharatnam–Berry Metasurface with Maximal Cross-Polarization Efficiency. *Adv. Mater.* **2015**, *27*, 1195–1200.

59. Luo, W.; Xiao, S.; He, Q.; Sun, S.; Zhou, L. Photonic Spin Hall Effect with Nearly 100% Efficiency. *Adv. Opt. Mater.* **2015**, *3*, 1102–1108.

60. Pancharatnam, S. Generalized theory of interference, and its applications. *Proc. Indian Acad. Sci. Sect.* **1956**, *44*, 247–262.

61. Berry, M.V. Quantal Phase Factors Accompanying Adiabatic Changes. *Proc. R. Soc. A Math. Phys. Eng. Sci.* **1984**, *392*, 45–57.

62. Bomzon, Z.; Kleiner, V.; Hasman, E. Pancharatnam–Berry phase in space-variant polarization-state manipulations with subwavelength gratings. *Opt. Lett.* **2001**, *26*, 1424–1426.

63. Bomzon, Z.; Biener, G.; Kleiner, V.; Hasman, E. Space-variant Pancharatnam–Berry phase optical elements with computer-generated subwavelength gratings. *Opt. Lett.* **2002**, *27*, 1141–1143.

64. Biener, G.; Niv, A.; Kleiner, V.; Hasman, E. Formation of helical beams by use of Pancharatnam–Berry phase optical elements. *Opt. Lett.* **2002**, *27*, 1875–1877.

65. Hasman, E.; Kleiner, V.; Biener, G.; Niv, A. Polarization dependent focusing lens by use of quantized Pancharatnam–Berry phase diffractive optics. *Appl. Phys. Lett.* **2003**, *82*, 328–330.

66. Menzel, C.; Rockstuhl, C.; Lederer, F. Advanced Jones calculus for the classification of periodic metamaterials. *Phys. Rev. A* **2010**, *82*, 053811.

67. Pors, A.; Nielsen, M.G.; Bozhevolnyi, S.I. Plasmonic metagratings for simultaneous determination of Stokes parameters. *Optica* **2015**, *2*, 716–723.

68. Pors, A.; Bozhevolnyi, S.I. Waveguide Metacouplers for In-Plane Polarimetry. *Phys. Rev. Appl.* **2016**, *5*, 064015.

69. Bomzon, Z.; Biener, G.; Kleiner, V.; Hasman, E. Spatial Fourier-transform polarimetry using space-variant subwavelength metal-stripe polarizers. *Opt. Lett.* **2001**, *26*, 1711–1713.

70. Wen, D.; Yue, F.; Kumar, S.; Ma, Y.; Chen, M.; Ren, X.; Kremer, P.E.; Gerardot, B.D.; Taghizadeh, M.R.; Buller, G.S.; et al. Metasurface for characterization of the polarization state of light. *Opt. Express* **2015**, *23*, 10272–10281.

71. Shaltout, A.; Liu, J.; Kildishev, A.; Shalaev, V. Photonic spin Hall effect in gap plasmon metasurfaces for on-chip chiroptical spectroscopy. *Optica* **2015**, *2*, 860–863.

72. Yin, X.; Ye, Z.; Rho, J.; Wang, Y.; Zhang, X. Photonic spin Hall effect at metasurfaces. *Science* **2013**, *339*, 1405–1407.

73. Wu, P.C.; Chen, J.W.; Yin, C.W.; Lai, Y.C.; Chung, T.L.; Liao, C.Y.; Chen, B.H.; Lee, K.W.; Chuang, C.J.; Wang, C.M.; et al. Visible Metasurfaces for On-chip Polarimetry. *ACS Photon.* **2017**. doi:10.1021/acsphotonics.7b01527.

74. Mueller, J.P.B.; Leosson, K.; Capasso, F. Ultracompact metasurface in-line polarimeter. *Optica* **2016**, *3*, 42–47.

75. Juhl, M.; Mendoza, C.; Mueller, J.B.; Capasso, F.; Leosson, K. Performance characteristics of 4-port in-plane and out-of-plane in-line metasurface polarimeters. *Opt. Express* **2017**, *25*, 28697–28709.

76. Wei, S.; Yang, Z.; Zhao, M. Design of ultracompact polarimeters based on dielectric metasurfaces. *Opt. Lett.* **2017**, *42*, 1580–1583.

77. Chen, W.T.; Török, P.; Foreman, M.R.; Liao, C.Y.; Tsai, W.Y.; Wu, P.R.; Tsai, D.P. Integrated plasmonic metasurfaces for spectropolarimetry. *Nanotechnology* **2016**, *27*, 224002.

78. Ding, F.; Pors, A.; Chen, Y.; Zenin, V.A.; Bozhevolnyi, S.I. Beam-Size-Invariant Spectropolarimeters Using Gap-Plasmon Metasurfaces. *ACS Photon.* **2017**, *4*, 943–949.

79. Maguid, E.; Yulevich, I.; Veksler, D.; Kleiner, V.; Brongersma, M.L.; Hasman, E. Photonic spin-controlled multifunctional shared-aperture antenna array. *Science* **2016**, *352*, 1202–1206.

80. Maguid, E.; Yulevich, I.; Yannai, M.; Kleiner, V.; Brongersma, M.L.; Hasman, E. Multifunctional Interleaved Geometric Phase Dielectric Metasurfaces. *Light Sci. Appl.* **2017**, *6*, e17027.

81. Khorasaninejad, M.; Chen, W.T.; Oh, J.; Capasso, F. Super-Dispersive Off-Axis Meta-Lenses for Compact High Resolution Spectroscopy. *Nano Lett.* **2016**, *16*, 3732–3737.

82. Zhu, A.Y.; Chen, W.T.; Khorasaninejad, M.; Oh, J.; Zaidi, A.; Mishra, I.; Devlin, R.C.; Capasso, F. Ultra-compact visible chiral spectrometer with meta-lenses. *APL Photon.* **2017**, *2*, 036103.

83. Chevalier, P.; Bouchon, P.; Jaeck, J.; Lauwick, D.; Bardou, N.; Kattnig, A.; Pardo, F.; Haïdar, R. Absorbing metasurface created by diffractionless disordered arrays of nanoantennas. *Appl. Phys. Lett.* **2015**, *107*, 251108.

84. Pisano, E.; Coello, V.; Garcia-Ortiz, C.E.; Chen, Y.; Beermann, J.; Bozhevolnyi, S.I. Plasmonic channel waveguides in random arrays of metallic nanoparticles. *Opt. Express* **2016**, *24*, 17080–17089.

85. Veksler, D.; Maguid, E.; Shitrit, N.; Ozeri, D.; Kleiner, V.; Hasman, E. Multiple wavefront shaping by metasurface based on mixed random antenna groups. *ACS Photon.* **2015**, *2*, 661–667.

86. Maguid, E.; Yannai, M.; Faerman, A.; Yulevich, I.; Kleiner, V.; Hasman, E. Disorder-induced optical transition from spin Hall to random Rashba effect. *Science* **2017**, *358*, 1411–1415.

87. Jang, M.; Horie, Y.; Shibukawa, A.; Brake, J.; Liu, Y.; Kamali, S.M.; Arbabi, A.; Ruan, H.; Faraon, A.; Yang, C. Wavefront shaping with disorder-engineered metasurfaces. *Nat. Photon.* **2018**, *12*, 84–90.

88. Chen, Y.; Ding, F.; Coello, V.; Bozhevolnyi, S.I. On-chip spectropolarimetry by fingerprinting with random surface arrays of nanoparticles. *ACS Photon.* **2018**, arXiv:1709.07617v1.

89. Allen, L.; Beijersbergen, M.W.; Spreeuw, R.; Woerdman, J. Orbital angular momentum of light and the transformation of Laguerre-Gaussian laser modes. *Phys. Rev. A* **1992**, *45*, 8185, doi:10.1103/PhysRevA.45.8185.

90. Bliokh, K.Y.; Rodríguez-Fortuño, F.; Nori, F.; Zayats, A.V. Spin-orbit interactions of light. *Nat. Photon.* **2015**, *9*, 796–808.

91. Genevet, P.; Yu, N.; Aieta, F.; Lin, J.; Kats, M.A.; Blanchard, R.; Scully, M.O.; Gaburro, Z.; Capasso, F. Ultra-thin plasmonic optical vortex plate based on phase discontinuities. *Appl. Phys. Lett.* **2012**, *100*, 013101.

92. Karimi, E.; Schulz, S.A.; De Leon, I.; Qassim, H.; Upham, J.; Boyd, R.W. Generating optical orbital angular momentum at visible wavelengths using a plasmonic metasurface. *Light Sci. Appl.* **2014**, *3*, e167.

93. Shalaev, M.I.; Sun, J.; Tsukernik, A.; Pandey, A.; Nikolskiy, K.; Litchinitser, N.M. High-efficiency all-dielectric metasurfaces for ultracompact beam manipulation in transmission mode. *Nano Lett.* **2015**, *15*, 6261–6266.

94. Yue, F.; Wen, D.; Zhang, C.; Gerardot, B.D.; Wang, W.; Zhang, S.; Chen, X. Multichannel Polarization-Controllable Superpositions of Orbital Angular Momentum States. *Adv. Mater.* **2017**, *29*, doi:10.1002/adma.201603838.

95. Afshinmanesh, F.; White Justin, S.; Cai, W.; Brongersma Mark, L. Measurement of the polarization state of light using an integrated plasmonic polarimeter. *Nanophotonics* **2012**, *1*, 125–129.

96. Li, W.; Coppens, Z.J.; Besteiro, L.V.; Wang, W.; Govorov, A.O.; Valentine, J. Circularly polarized light detection with hot electrons in chiral plasmonic metamaterials. *Nat. Commun.* **2015**, *6*, 8379, doi:10.1038/ncomms9379.

97. Gorodetski, Y.; Shitrit, N.; Bretner, I.; Kleiner, V.; Hasman, E. Observation of optical spin symmetry breaking in nanoapertures. *Nano Lett.* **2009**, *9*, 3016–3019.

98. Chen, W.; Abeysinghe, D.C.; Nelson, R.L.; Zhan, Q. Experimental confirmation of miniature spiral plasmonic lens as a circular polarization analyzer. *Nano Lett.* **2010**, *10*, 2075–2079.

99. Li, Q.T.; Dong, F.; Wang, B.; Chu, W.; Gong, Q.; Brongersma, M.L.; Li, Y. Free-space optical beam tapping with an all-silica metasurface. *ACS Photon.* **2017**, *4*, 2544–2549.

100. Ahn, S.H.; Guo, L.J. Large-area roll-to-roll and roll-to-plate nanoimprint lithography: A step toward high-throughput application of continuous nanoimprinting. *ACS Nano* **2009**, *3*, 2304–2310.

applied
sciences

MDPI

Article

Measurement Matrix Analysis and Radiation Improvement of a Metamaterial Aperture Antenna for Coherent Computational Imaging

Na Kou, Long Li *, Shuncheng Tian and Yuanchang Li

Key Laboratory of High Speed Circuit Design and EMC of Ministry of Education, School of Electronic Engineering, Collaborative Innovation Center of Information Sensing and Understanding, Xidian University, Xi'an 710071, China; kouna02091322@126.com (N.K.); sctian@xidian.edu.cn (S.T.); ycli@stu.xidian.edu.cn (Y.L.)
* Correspondence: lilong@mail.xidian.edu.cn; Tel.: +86-29-8820-1157

Received: 26 July 2017; Accepted: 7 September 2017; Published: 12 September 2017

Abstract: A metamaterial aperture antenna (MAA) that generates frequency-diverse radiation field patterns has been introduced in the context of microwave wave imaging to perform compressive image reconstruction. This paper presents a new metamateriapl aperture design, which includes two kinds of metamaterial elements with random distribution. One is a high-Q resonant element whose resonant frequency is agile, and the other one is a low-Q element that has a high radiation efficiency across frequency band. Numerical simulations and measurements show that the radiation efficiency of up to 60% can be achieved for the MAA and the far-field patterns owns good orthogonality, when using the complementary electric-field-coupled (CELC) element and the complementary Jerusalem cross (CJC) element with a random distribution ratio of 4 to 1, which could be effectively used to reconstruct the target scattering scene.

Keywords: agile frequency response; compressive image reconstruction; metamaterial aperture antenna

---

1. Introduction

In recent years, several physical platforms based on metamaterial aperture antennas (MAAs) have been used to realize computational imaging, as shown in Figure 1. In general, all natural scenes can be compressed on some basis. Scenes can be perfectly reconstructed with significantly fewer measurement modes than the space bandwidth product (SBP) [1–3]. These measurement modes are composed of radiated field patterns of the metamaterial aperture antenna (MAA) at different frequencies. For image reconstruction schemes, which use an arbitrary set of measurement modes, it is essential that the modes are as orthogonal to each other as possible [4] so that the correlation of the modes or field patterns is small [5–7]. Since the measurement modes are indexed by frequency, the low correlation of field patterns can be interpreted as far-field patterns with strong frequency diversity. Hence, to obtain a metamaterial aperture that generates the frequency-agile far-field patterns that are as orthogonal as possible, elements distributed on the aperture must have a strong resonance with a high Q-value. Other approaches to frequency-diverse imaging have also been pursued, including multiply scattering structures, such as mode-mixing cavities and dynamic metamaterial apertures [8–10]. Fractal models have been used to design fractal antennas with very special properties: about one-tenth of a wavelength and a pre-fractal geometrical configuration [11] could be used in metamaterial miniaturized technology, which is needed for coherent imaging. The fractional signal, especially its geometrical interpretation, which gives a powerful mathematical tool to model the most advanced concepts of modern physics, could help us to build a simpler mathematical model of the relationship between the metamaterial aperture antenna and its radiation feature, which will improve the speed of full-wave simulation [12]. The initial metamaterial aperture which consists of numerous subwavelength-resonant radiators present a trade-off between the

Q factor and radiation efficiency (each resonator distributed on aperture is assigned a resonant frequency randomly selected from the bandwidth of operation) [13,14], and most of the power is radiated by the elements closest to the feed, resulting in poor aperture efficiency.

In this paper, the correlation of far-field patterns for metamaterial apertures with different modulations is analyzed, and a trade-off strategy between orthogonality of the far-field patterns and radiation efficiency of the MAA is proposed. The metamaterial aperture is composed of complementary electric-field-coupled elements (CELCs) [15] and complementary Jerusalem cross (CJC) [16] unit cells. The CELC element is strongly resonant with frequency diversity, while the CJC unit cell has high radiation efficiency across the frequency band. By analyzing the filling ratio of the two different elements on the MAA, a relatively low correlation of the far-field patterns and high radiation efficiency is obtained. MAAs with a 20 × 20 randomly distributed elements array are theoretically analyzed in the frequency band from 33 to 37 GHz. Numerical simulations show that the radiation efficiency of 60% can be achieved, which yields improved SNRs when using the complementary electric-field-coupled (CELC) element and the CJC element with a random distribution ratio of 4 to 1 on the aperture. A prototype of the MAA with 120 × 120 randomly distributed elements whose total size is 250 mm × 250 mm × 0.5 mm is fabricated to verify the effectiveness of the design and analysis procedure, and to show that the measured far-field patterns across the frequency band from 33 to 37 GHz could be used to effectively reconstruct the target scene with scatterers.

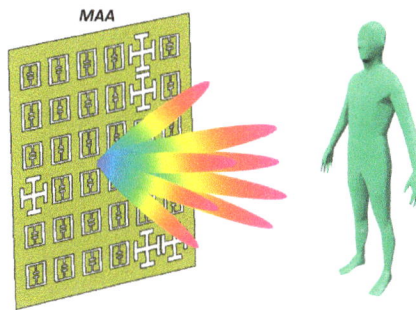

Figure 1. A metamaterial aperture illuminating a human target scene.

2. The Calculation for the Correlation of Far-Field Patterns of an MAA

In the context of compressed sensing, the measurement modes can be selected as the radiated far-field patterns of the MAA in a given frequency range. Consider the measurement matrix H, which corresponds to the frequency-diverse far-field patterns. H is an $M \times N$ matrix. M represents the quantity of the measurement modes, which is equal to the number of the sampled frequency points. N denotes the far-field pattern pixels at one frequency point, which is the product of the azimuth angles and elevation angles. The matrix reconstruction metric is the average mutual coherence μ_g, which is defined as follows [13].

Firstly, with the measurement matrix, the Gram matrix can be obtained [17,18]:

$$G = H_n{}^\dagger H_n \tag{1}$$

where the superscript † represents the conjugate transpose of the matrix, and H_n is the measurement matrix H with normalized columns. The average mutual coherence is

$$\mu_g = \frac{\sum\limits_{i \neq j} |G_{ij}|}{N(N-1)} \tag{2}$$

As pointed out by Lipworth et al. [13], an increase in the Q-factor of the element on the metamaterial aperture causes μ_g to decrease, which indicates that less correlation between measurements represents better orthogonality of the far-field patterns. Here, the frequency quality factor Q can be defined as

$$Q = \frac{f_0}{2\Delta f} \qquad (3)$$

where f_0 is the center operating frequency, and $2\Delta f$ represents the frequency range, which the amplitude response decreases by 3 dB. Usually, when the operating frequency band is given, the center frequency f_0 is fixed and the frequency interval Δf could be changed by selecting different resonant elements with varying frequency agility features. For the strongly resonant element with a high Q-value, its frequency selectivity is good, which means the Δf or the bandwidth is relatively small. By using this kind of element on the MAA, frequency-agile field patterns with good orthogonality can be obtained. As a result, the target scene can be perfectly reconstructed by utilizing a simple metamaterial aperture antenna instead of mechanical scanning or antenna arrays.

Figure 2 shows the flow chart for calculation procedure of μ_g. Firstly, an MAA with randomly distributed elements is designed. At each frequency point, the phase and amplitude responses of the elements on the MAA can be obtained by numerical simulations. Next, the far-field patterns across the frequency band can be calculated using array synthesis technology. Finally, far-field patterns form the measurement matrix and the average mutual coherence μ_g can be obtained. Here we will focus on the theoretical analysis for the variation of μ_g when the Q-factor of the resonant elements of the MAA changes.

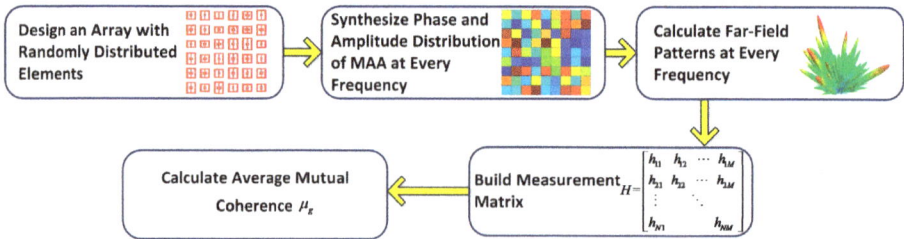

Figure 2. Flow chart of the μ_g analysis procedure.

3. An Analysis of Average Mutual Coherence μ_g

In order to obtain a set of frequency-diverse far-field patterns whose average mutual coherence μ_g is low, it is essential that different amplitude and phase distributions on one metamaterial aperture are realized at different frequencies from the point of antenna design. However, the strongly resonant elements with frequency-agile responses both on amplitude and phase responses are not realistically achievable. Hence, the phase and amplitude modulations of MAA are performed separately to analyze the average mutual coherence μ_g of its far-field patterns.

For the phase modulation of MAA, the phase distribution on the aperture antenna varies when frequency changes while the amplitude distribution on MAA is uniform. Here, 20 × 20 resonant elements with varying phase responses at different frequencies are randomly distributed on the metamaterial aperture antenna, which is analyzed by using array synthesis technology in the frequency band from 33 to 37 GHz. The average mutual coherence μ_g of the far-field patterns of the MAA is calculated using elements with varying Q-values (or Δf). When the Δf of the elements is changed from 0.005 to 0.5 GHz, the calculated μ_g varies from 0.02 to 0.13, as shown in Figure 3.

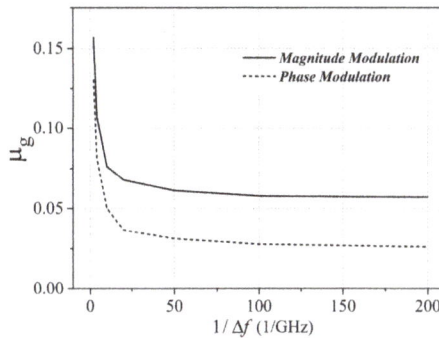

Figure 3. Comparison of the average mutual coherence μ_g between phase and amplitude modulation.

Similarly, for the amplitude modulation of MAA, the amplitude distribution on the aperture is diverse at different frequencies, while the phase distribution on MAA is constant. And 20 × 20 resonant elements with diverse amplitude responses in frequency are also distributed on the aperture randomly. By changing Δf, the average mutual coherence μ_g of the far-field patterns of the MAA is calculated. When the bandwidth Δf of the element is changed from 0.005 to 0.5 GHz, the calculated μ_g varies from 0.05 to 0.15. The comparison of μ_g between the phase and amplitude modulation is shown in Figure 3. It can be seen from Figure 3 that the calculated μ_g of phase modulation is slightly smaller than that of amplitude modulation. When the bandwidth of the element becomes smaller, the average mutual coherence becomes lower. However, the average mutual coherence becomes flat when Δf is less than 0.005 GHz. It is worth mentioning that the average mutual coherence is irrelevant with respect to the specific location of the center frequency, but it is relative to Δf. It is common that, at higher operating frequencies, such as the millimeter-wave frequency band, the resonant element often has a larger Δf (usually ≥0.1 GHz). Hence, it is much more difficult to design a resonant element with $\Delta f \leq 0.1$ GHz at the millimeter-wave band, such as 35 GHz.

4. A Method of Improving the Radiation Efficiency of an MAA

In general, resonant elements with varying amplitude responses in frequency are realistically achievable. Specifically, different amplitude distributions of the elements on an MAA are represented by the varying radiation efficiencies on the aperture. However, there is a trade-off between the Q-factor and the radiation efficiency of these elements [19]. Additionally, an increase in the Q-factor of the element on the metamaterial aperture causes μ_g to decrease [13]. Hence, a trade-off between the average mutual coherence μ_g and the radiation efficiency of the MAA must be made.

Usually the MAA is composed of all the strongly resonant elements, such as a CELC [15]. Figure 4 shows the radiation characteristic of a CELC element. The PMC (Perfect Magnetic Conductor) boundary used here is meant to represent the periodic boundary condition that is used to model an infinite metamaterial aperture antenna in practice. It can be seen that, when size parameters g and l decrease, the resonant frequency of the CELC increases and the CELC element has an agile radiation efficiency response across the frequency band. However, the element has a very low radiation efficiency, with a peak value of approximate 50% and a bandwidth of $\Delta f \approx 0.1$ GHz. In this case, a low Q-factor element can be introduced to increase the radiation efficiency without significantly destroying μ_g. The CJC unit cell [16] is adopted, which has a low quality factor but a radiation efficiency that is superior to that of the CELC element, as shown in Figure 5.

In this paper, the improved strategy of the MAA design is a combination of CELC and CJC elements. However, the ratio of the CJC elements distributed on the aperture is crucial, which causes a balance problem between the μ_g and the radiation efficiency of the antenna. To illustrate this, we analyze the performances of the MAA with 20 × 20 elements, when the ratio of CJC elements changes from 0 to 50%. On the aperture, the CELC elements are used to generate frequency-diverse field patterns. The CJC elements that are used to improve the radiation efficiency are placed on the remaining part of the aperture with a certain ratio. Using a full-wave simulation tool, the radiation efficiency of the CJC element is obtained in the frequency band from 33 to 37 GHz, as shown in Figure 5b. It can be seen that the CJC element has a good radiation ability.

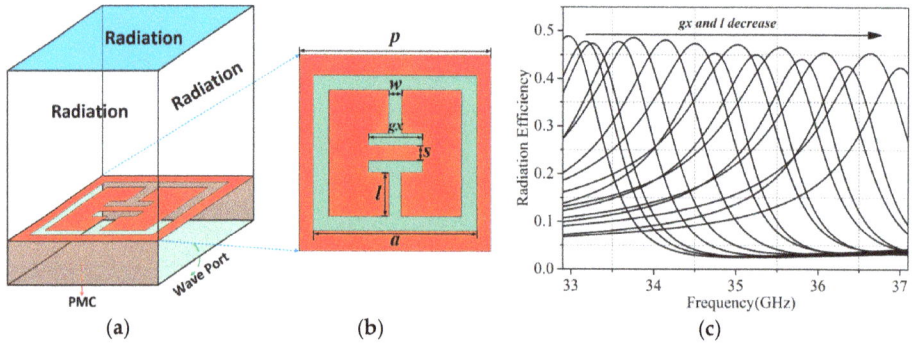

Figure 4. The complementary electric-field-coupled (CELC) element with a unit cell size of 2 mm: (**a**) full-wave simulation model; (**b**) its top view; (**c**) the variation in radiation efficiency when *l* and *g* change.

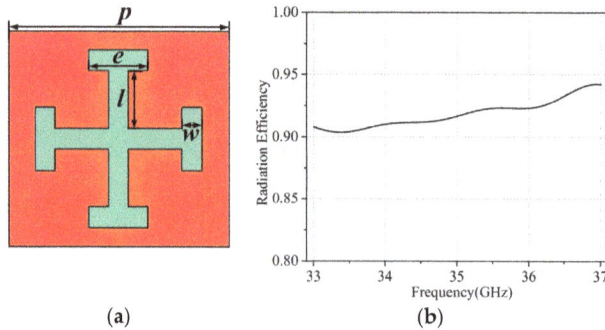

Figure 5. (**a**) Top view of complementary Jerusalem cross (CJC) element with a unit cell size of 2 mm and (**b**) a radiation efficiency of the element.

When the filling ratio of CJC elements on the MAA is changed from 0 to 50%, as shown in Figure 6, the average mutual coherence μ_g will vary from 0.115 to 0.377, as shown in Figure 7a. The radiation efficiency of the MAAs with different ratios of CJC elements is shown in Figure 7b. We can see in Figure 7 that the increase in radiation efficiency causes μ_g to increase. As a result, to obtain a relatively high radiation efficiency and a low μ_g, the ratio of CJC elements must be selected as 20% since its radiation efficiency has been improved to around 60% and its capability of imaging reconstruction approaches that of a metamaterial aperture without CJC elements.

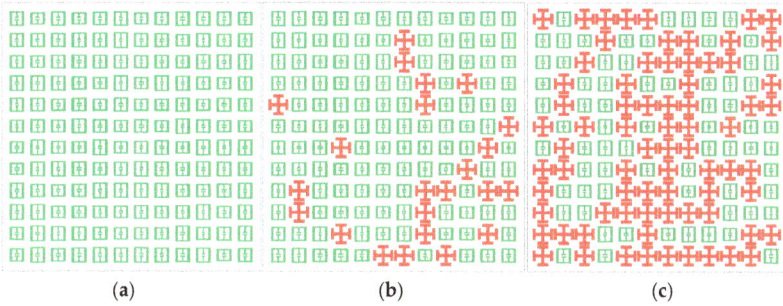

Figure 6. The schematic of simulated metamaterial aperture antenna (MAA) with different ratios of CJC elements: (**a**) 0%; (**b**) 20%; (**c**) 50%.

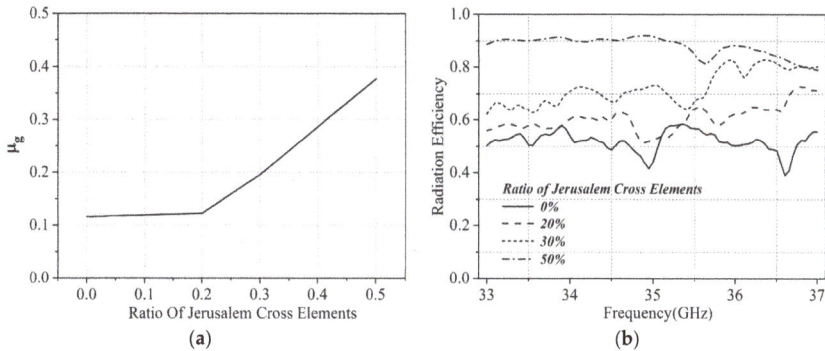

Figure 7. (**a**) Calculated average mutual coherence μ_g and (**b**) the radiation efficiency of different MAAs with different filling ratios of CJC elements.

Thus, the prototype of a metamaterial aperture antenna, with 120×120 randomly distributed CELC and CJC elements, whose total size is 250 mm $\times$ 250 mm $\times$ 0.5 mm was fabricated, measured, and analyzed. The fabricated MAA consisted of 80% CELC elements and 20% CJC elements, as shown in Figure 8b,c. The feed structure of the metamaterial aperture antenna is shown in Figure 8a below. The array consists of four uniform feed structures. The feed structure contains one impedance matching capacitor and four impedance matching inductive pins, and a coaxial feed pin is at the center of the structure. The measurement was performed using near-field planar-scanning techniques and near-to-far-field transformation [20]. The measured far-field patterns across the frequency band from 33 to 37 GHz are shown in Table 1. It can be seen that the designed MAA has frequency-diverse far-field patterns.

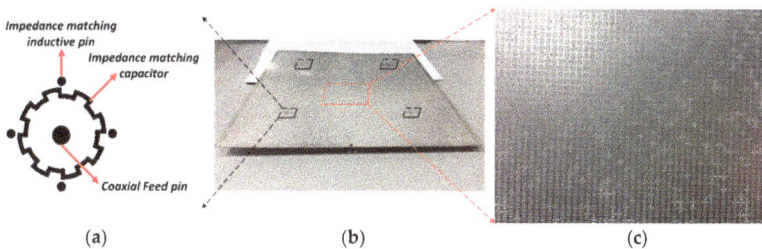

Figure 8. Geometry of the fabricated MAA: (**a**) feed structure (the black capacitor part is dielectric, and the remaining part is copper); (**b**) total view; (**c**) local view.

Table 1. Measured field patterns of the metamaterial aperture antenna at different frequencies.

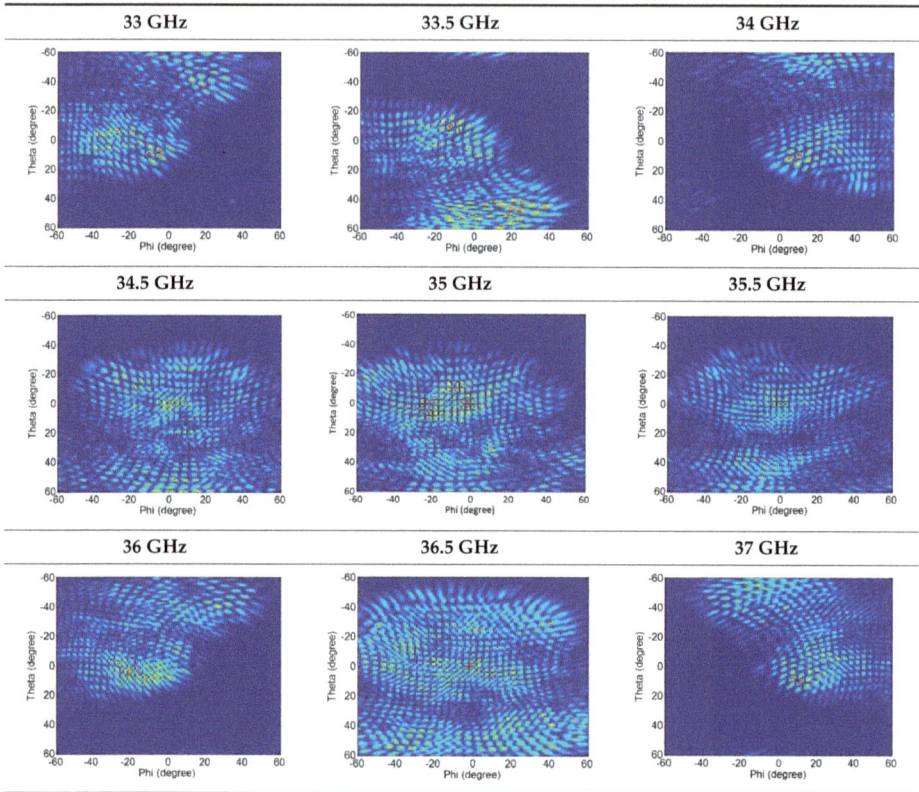

33 GHz	33.5 GHz	34 GHz

34.5 GHz	35 GHz	35.5 GHz

36 GHz	36.5 GHz	37 GHz

5. Discussion

The capability of imaging reconstruction of the fabricated metamaterial aperture antenna was explored based on the measured data. The specific parameters of the MAA's coherent computational imaging system are shown in Table 2. Figure 9a shows the original sparse target scene, which includes two scatterers, and Figure 9b shows this scene reconstructed via the pseudo-inverse method. It can be seen in Figure 9 that the designed MAA could be effectively used to reconstruct the target scenes. In conclusion, when using the complementary electric-field-coupled (CELC) element and the complementary Jerusalem cross (CJC) element with a random distribution ratio of 4 to 1, the radiation efficiency of the MAA can achieve 60%, and the far-field patterns has good orthogonality, which could be effectively used to reconstruct the target scatterers scene.

Table 2. System parameters of the metamaterial aperture antenna.

Parameters	Value
Field of View(Elevation)	-60–$60°$
Frequency sampling interval	0.1 GHz
Elevation sampling interval	$1°$
Azimuth sampling interval	$1°$

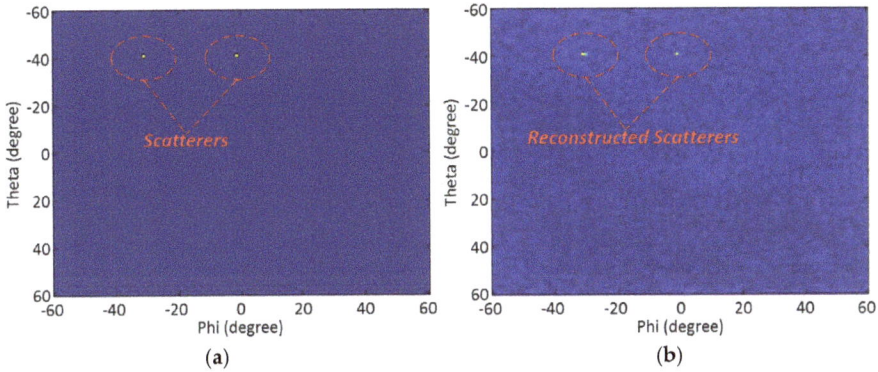

Figure 9. Target scenes: (**a**) true target scene consisting of two point scatterers; (**b**) the same scene reconstructed by the fabricated MAA.

Since the capability of imaging reconstruction of the fabricated metamaterial aperture antenna is explored based on the measured far-field data, the range of the target is in the far-field region. In this case, since the dimension of the fabricated metamaterial aperture antenna is 250 mm × 250 mm × 0.5 mm, and the center frequency is 35 GHz, the far-field region is $r \geq (2 \times 0.25 \text{ m} \times 0.25 \text{ m})/\lambda_0 = 14.5$ m (λ_0 is the wavelength at the center frequency in free space). As a result, the true target is assumed to be 15 m away from the metamaterial aperture antenna.

The capability of imaging reconstruction of the new proposed metamaterial aperture antenna was determined based on simulated targets at the present stage [21]. As for larger targets, we also employed a simulated metal sheet target to validate the capability of imaging reconstruction of the new proposed MAA, as shown in Figure 10 below. In Figure 10, we can see that the measured far-field data could be used to reconstruct the location and shape of the metal sheet more precisely, but not perfectly. Going forward, more practical experiments with a more precise reconstruction of larger targets will be crucial.

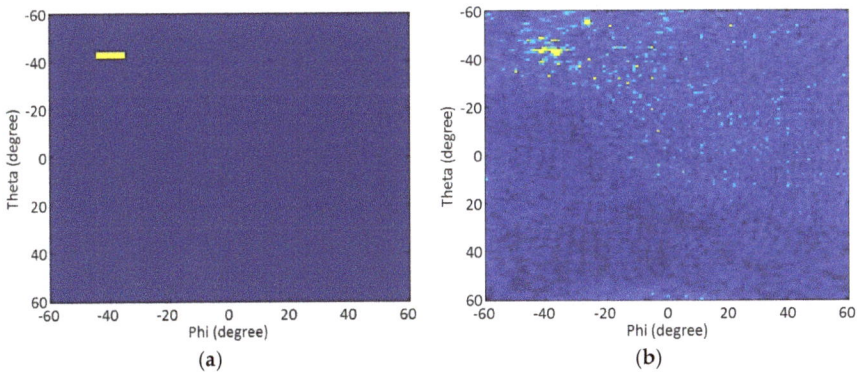

Figure 10. Target scenes: (**a**) true target scene consisting of one metal sheet scatterers; (**b**) the same scene reconstructed by the fabricated MAA.

6. Conclusions

A new metamaterial aperture antenna (MAA) composed of two kinds of metamaterial elements with random distribution was proposed, designed, and measured to perform compressive image

Appl. Sci. **2017**, *7*, 933

reconstruction. One element has a high *Q*-value with an agile frequency response, and the other element has a relatively low *Q*-factor and has a high radiation efficiency. The orthogonality of the far-field patterns and the improvement in the radiation efficiency of the MAA were theoretically analyzed. Numerical simulations and measurements show that the radiation efficiency of the MAA can achieve 60%, and the far-field patterns have good orthogonality; the MAA can therefore be effectively used to reconstruct scenes with target scatterers when using the complementary electric-field-coupled (CELC) element and the complementary Jerusalem cross (CJC) element with a random distribution ratio of 4 to 1. Moreover, the proposed MAA can be effectively used for coherent computational imaging.

Acknowledgments: This work was supported by the National Natural Science Foundation of China under Contract No. 51477126, the Technology Explorer and Innovation Research Project, and Fundamental Research Funds for the Central Universities (K5051202051 and SPSZ021409).

Author Contributions: Na Kou conceived the idea of improving the radiation efficiency of the metamaterial aperture antenna and performed the simulations and experiments; Long Li provided the main instructions of the ideas and experiments; Shuncheng Tian and Yuanchang Li designed the feed of the antenna and performed the experiments.

Conflicts of Interest: The authors declare no conflict of interest. The founding sponsors had no role in the design of the study; in the collection, analyses, or interpretation of data; in the writing of the manuscript; or in the decision to publish the results.

References

1. Hunt, J.; Driscoll, T.; Mrozack, A.; Lipworth, G.; Reynolds, M.; Brady, D.; Smith, D.R. Metamaterial apertures for computational imaging. *Science* **2013**, *339*, 310–313. [CrossRef] [PubMed]
2. Brady, D.J.; Choi, K.; Marks, D.L.; Horisaki, R.; Lim, S. Compressive holography. *Opt. Express* **2009**, *17*, 13040–13049. [CrossRef] [PubMed]
3. Cull, C.F.; Wikner, D.A.; Mait, J.N.; Mattheiss, M.; Brady, D.J. Millimeter-wave compressive holography. *Appl. Opt.* **2010**, *49*, 67–82. [CrossRef] [PubMed]
4. Lohmann, A.W.; Testorf, M.E.; Ojeda-Castañeda, J. The space-bandwidth product, applied to spatial filtering and to holography. *Med. J. Aust.* **2006**, *1*, 565.
5. Marks, D.L.; Gollub, J.; Smith, D.R. Spatially resolving antenna arrays using frequency diversity. *J. Opt. Soc. Am. A* **2016**, *33*, 899–912. [CrossRef] [PubMed]
6. Sleasman, T.; Imani, M.F.; Gollub, J.N.; Smith, D.R. Microwave Imaging Using a Disordered Cavity with a Dynamically Tunable Impedance Surface. *Phys. Rev. Appl.* **2016**, *6*, 054019. [CrossRef]
7. Sleasman, T.; Boyarsk, M.; Imani, M.F.; Gollub, J.N.; Smith, D.R. Design considerations for a dynamic metamaterial aperture for computational imaging at microwave frequencies. *J. Opt. Soc. Am. B* **2016**, *33*, 1098. [CrossRef]
8. Fromenteze, T.; Yurduseven, O.; Imani, M.F.; Gollub, J.; Decroze, C.; Carsenat, D.; Smith, D.R. Computational imaging using a mode-mixing cavity at microwave frequencies. *Appl. Phys. Lett.* **2015**, *106*, 194104. [CrossRef]
9. Fromenteze, T.; Decroze, C.; Carsenat, D. Waveform Coding for Passive Multiplexing: Application to Microwave Imaging. *IEEE Trans. Antennas Propag.* **2014**, *63*, 593–600. [CrossRef]
10. Sleasman, T.; Imani, M.F.; Gollub, J.N.; Smith, D.R. Dynamic metamaterial aperture for microwave imaging. *Appl. Phys. Lett.* **2015**, *107*, 204104. [CrossRef]
11. Guariglia, E. Entropy and Fractal Antennas. *Entropy* **2016**, *18*, 84. [CrossRef]
12. Guariglia, E. Fractional Derivative of the Riemann Zeta Function. *Fract. Dyn.* **2015**, *21*, 357–368.
13. Lipworth, G.; Mrozack, A.; Hunt, J.; Marks, D.L.; Driscoll, T.; Brady, D.; Smith, D.R. Metamaterial microwave holographic imaging system. *J. Opt. Soc. Am. A* **2013**, *30*, 1603. [CrossRef] [PubMed]
14. Lipworth, G.; Rose, A.; Yurduseven, O.; Gowda, V.R.; Imani, M.F.; Odabasi, H.; Trofatter, P.; Gollub, J.; Smith, D.R. Comprehensive simulation platform for a metamaterial imaging system. *Appl. Opt.* **2015**, *54*, 9343–9353. [CrossRef] [PubMed]
15. Schurig, D.; Mock, J.J.; Smith, D.R. Electric-field-coupled resonators for negative permittivity metamaterials. *Appl. Phys. Lett.* **2006**, *88*, 041109. [CrossRef]
16. Arnaud, J.A.; Pelow, F.A. Resonant-Grid Quasi-Optical Diplexers. *Electron. Lett.* **1973**, *9*, 589–590. [CrossRef]

17. Duarte-Carvajalino, J.M.; Sapiro, G. Learning to sense sparse signals: Simultaneous sensing matrix and sparsifying dictionary optimization. *IEEE Trans. Image Process.* **2009**, *18*, 1395–1408. [CrossRef] [PubMed]
18. Elad, M. Optimized Projections for Compressed Sensing. *IEEE Trans. Signal Process.* **2007**, *55*, 5695–5702. [CrossRef]
19. Chu, L.J. Physical Limitations of Omni-Directional Antennas. *J. Appl. Phys.* **1948**, *19*, 1163–1175. [CrossRef]
20. Balanis, C.A. *Antenna Theory: Analysis and Design*, 3rd ed.; Wiley-Interscience: Hoboken, NJ, USA; Harper & Row: Bel Air, CA, USA, 2005; pp. 1014–1026.
21. Wu, Z.; Zhang, L.; Liu, H.; Kou, N. Enhancing Microwave Metamaterial Aperture Radar Imaging Performance with Rotation Synthesis. *IEEE Sens. J.* **2016**, *16*, 8035–8043. [CrossRef]

applied
sciences

MDPI

Article

A Transplantable Frequency Selective Metasurface for High-Order Harmonic Suppression

Na Kou, Haixia Liu * and Long Li *

Key Laboratory of High Speed Circuit Design and EMC of Ministry of Education, School of Electronic Engineering, Collaborative Innovation Center of Information Sensing and Understanding, Xidian University, Xi'an 710071, China; kouna02091322@126.com
* Correspondence: hxliu@xidian.edu.cn (H.L.); lilong@mail.xidian.edu.cn (L.L.); Tel.: +86-029-8820-4458 (H.L.); +86-029-8820-1157 (L.L.)

Received: 5 November 2017; Accepted: 23 November 2017; Published: 1 December 2017

Abstract: A transplantable frequency selective metasurface element (FSMSE) for high-order harmonic suppression (HS) is presented in this paper. The proposed harmonic free FSMSE can be integrated with arbitrary frequency selective surfaces (FSSs) operating at the same frequency band. The designed HS-FSMSE is applied to two different types of FSSs for verification of harmonic suppression, respectively. One is a multilayer sub-wavelength patch-grid FSS that has weak resonant behavior, and the other is a complementary resonant loop FSS which has strong resonant behavior, especially for high-order harmonic waves. By integrating with HS-FSMSE, the two kinds of FSSs operating at 10 GHz are free of harmonic transmission bands up to 30 GHz. The simulation and measurement results show feasibility of the harmonic suppression FSMSW and good polarization and angle stabilities.

Keywords: frequency selective metasurface element (FSMSE); harmonic suppression; transplantable

---

1. Introduction

Metasurfaces have been widely applied in manipulation of electromagnetic wave, such as radar radomes [1], spatial filters [2], reflect arrays [3], artificial impedance surfaces [4], polarization converters [5], and absorbing materials [6], etc., because of their properties of allowing uninhibited transmission of electromagnetic waves in specific frequency bands while suppressing transmission in other undesired bands. In general, a band-pass spatial filter based on metasurface is also called frequency selective surface (FSS) which is transparent within the operation frequency band of the antenna and opaque at other frequencies. However, most FSSs have multiple spurious transmission windows in higher operating frequency bands. And the high frequency harmonic phenomenon will severely interfere with other antennas and radars which operate at the h-armonic frequency bands especially on the low-observable or stealth platforms. As a result, the FSSs with harmonic suppression feature are important for practical applications.

In [7–9], band-pass frequency selective surfaces with stop-band characteristics are reported. They use artificial absorbing coating, resistive high-impedance surface, and other techniques to achieve the harmonic absorbing properties. However, the FSSs in these designs are specific which means the technique of these harmonic suppression FSSs cannot be independent and transplanted. In this paper, a transplantable frequency selective metasurface element for harmonic suppression is designed. The frequency selective metasurface element (FSMSE) are integrated with two different types of band-pass FSSs to remove the spurious transmission windows outside of the desired operational pass-band, respectively. Full-wave simulations and experiments show feasibility of the harmonic suppression FSMSE, and measured results agree well with simulated ones.

2. Design of the Harmonic-Suppression Frequency Selective Metasurface Element (FSMSE)

In general, a FSS with band-pass feature intrinsically has spurious transmission windows outside of the desired operational pass-band. For an example, a classic non-resonant FSS [7] which consists of square metallic patch with fishnet grids in multilayer structure is shown in Figure 1a. Full-wave analyses show that the pass-band characteristic of the FSS is operating at about 10 GHz, but with harmonic transmission bands around 25 GHz, as shown in Figure 2. When the plane wave is oblique incidence with Transverse Electric (TE) and Transverse Magnetic (TM) polarizations, the spurious transmission windows become much more. Similarly, a typical resonant type FSS which is composed of complementary loop structure [10] is shown in Figure 1b, which also operates at 10 GHz. Because the FSS is resonant structure, it has several harmonic pass bands at the higher frequency bands outside of the main pass-band, as shown in Figure 3. The geometry parameters of the two FSSs are shown in Table 1. Thus it can be seen that the harmonic suppression technique should be adopted. In this paper, a transplantable FSMSE is proposed that can be integrated with different types of band-pass FSS structures without deteriorating the operating frequency of the pass band.

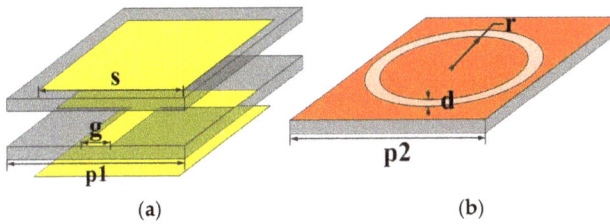

Figure 1. (a) Subwavelength multilayer patch and grid frequency selective surface (FSS), (b) resonant complementary loop FSS (The grey represents substrates, the yellow and orange parts represent metal).

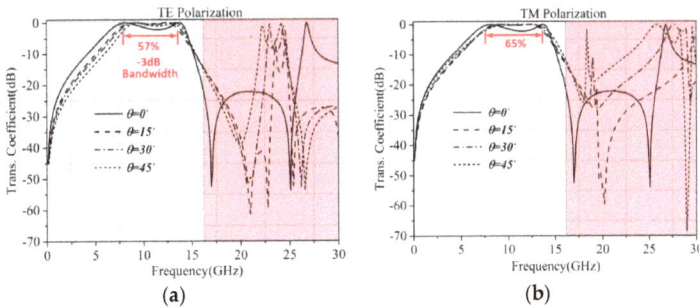

Figure 2. Transmission coefficient characteristics for oblique incidences on the non-resonant FSS for (a) Transverse Electric (TE) polarization and (b) Transverse Magnetic (TM) polarization (The red shadow represents harmonic transmission bands).

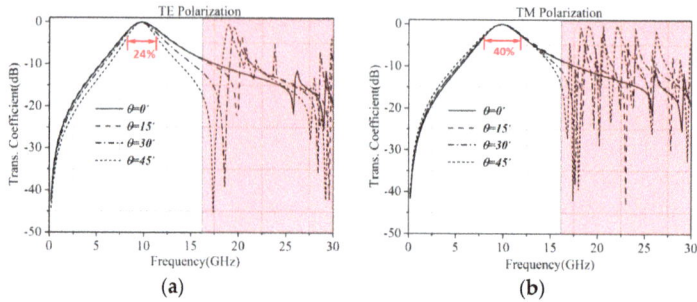

Figure 3. Transmission coefficient characteristics for oblique incidences on the resonant complementary loop FSS for (**a**) TE polarization and (**b**) TM polarization (The red shadow represents harmonic transmission bands).

Table 1. Parameters of the frequency selective surfaces (FSSs) shown in Figure 1.

Parameters	s	g	p1	r	d	p2
mm	5.5	0.5	6.6	4.3	0.5	10

The topology of the FSMSE for harmonic suppression is shown in Figure 4, which is composed of two-layer square loop elements that intrinsically own band-stop feature. It is known that the single-layer square loop FSMSE has a narrow stop band whose operating frequency could be adjusted by changing the length of square loop. As a result, when we want to obtain a relatively wide stop band, the multilayer square loop FSMSE with air spacing can be designed. At the same time, the pass band of 10 GHz has to be guaranteed. The equivalent circuit model is used to analyze the feature of the FSMSE for harmonic suppression, as shown in Figure 5.

Here the square loop element can be equivalent to a band stop circuit. As a result, the equivalent circuit of the two-layer square loop FSMSE is a two-stage cascaded circuit which is connected by a uniform transmission line whose length depends on the electric size of the air spacing, as shown in Figure 5. The geometry parameters are shown in Table 2. The lumped elements in the equivalent circuit can be obtained using bisquare fitting algorithm [11]. According to the equivalent circuits of each layer of the square loop FSMSE, the lumped parameters are extracted as follows: $L_{1s} = L_{1s1} = 0.0834$ nH, $L_{2s} = L_{2s1} = 0.001$ nH, $C_{2s} = 0.6677$ pF, $C_{2s1} = 0.4462$ pF, $L_{3s} = 0.0762$ nH, $L_{3s1} = 0.0624$ nH, $C_{3s} = 0.1834$ pF and $C_{3s1} = 0.2167$ pF. With these lumped elements and the circuit prototype, the optimal air spacing h could be obtained using circuit optimization. It can be seen that when h equals to 2 mm, the maximum stop-band range can be obtained. Figure 6 depicts the comparison of transmission coefficient of the harmonic suppression FSMSE between the full-wave simulation and the equivalent circuit, which shows good agreement with each other. Furthermore, the oblique incidences of TE and TM polarizations with different incident angles on the harmonic suppression FSMSE are analyzed, respectively, as shown in Figure 7. It can be seen that the proposed FSMSE has good polarization and angle stability. And the −20 dB suppression bandwidth is from 16.8 to 26.8 GHz.

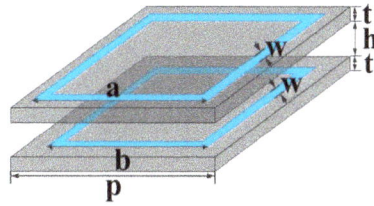

Figure 4. A 3D topology of harmonic suppression frequency selective metasurface element (FSMSE) (The blue part represents metal and the grey represents substrates).

Table 2. Parameters of the FSSs shown in Figure 4.

Parameters	a	b	w	p	t	h
mm	3	2.8	0.2	3.25	0.5	2

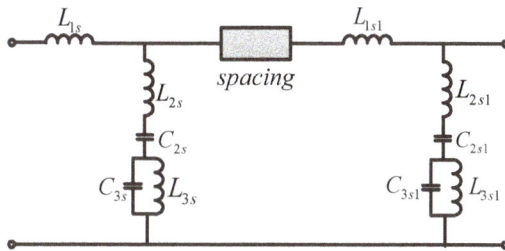

Figure 5. The equivalent circuit of harmonic suppression metasurface element.

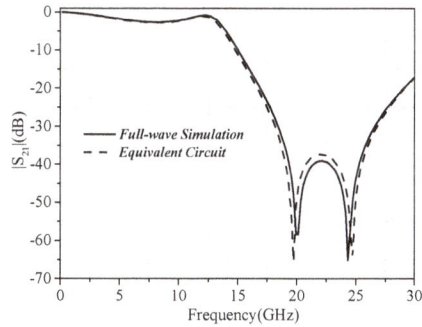

Figure 6. Comparison of transmission coefficients of the full-wave simulation and equivalent circuits of the harmonic suppression FSMSE.

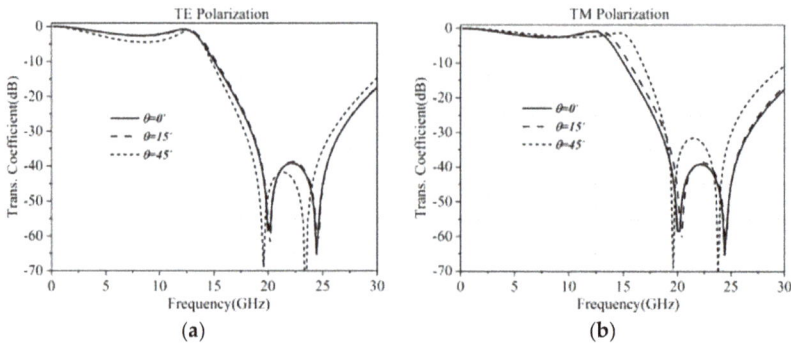

Figure 7. Transmission coefficients for oblique incidences with different angles on the harmonic suppression (HS) FSMSE, (**a**) TE polarization and (**b**) TM polarization.

Next, the designed harmonic suppression FSMSE can be integrated with the aforementioned non-resonant patch-grid FSS and the resonant complementary loop FSS, respectively, as shown in Figures 8 and 9. The air spacing h1 between the two original FSSs and the HS-FSMSE was also optimized using circuit analysis where the two types of FSSs are equivalent to band-pass circuits whose lumped elements can also be extracted by bisquare fitting algorithm. When h1 is 3 mm, the harmonic suppression feature can perfectly be achieved without deteriorating the operating frequency of the pass band. Figure 10 shows the transmission coefficient of non-resonant patch-grid FSS with HS-FSMSE for different incidence angles with TE and TM polarizations. It can be seen that the high-order harmonic waves can be suppressed below −40 dB after integrating with HS-FSMSE structure. Similarly, Figure 11 depicts the transmission coefficient of the resonant complementary loop FSS with HS-FSMSE for different incidence angles with TE and TM polarizations. The results show that the high-order harmonic waves can also be well suppressed even with the strong resonant behavior. In the meantime, we can see that the two different types of FSSs with same HS-FSMSE own good polarization and angle stabilities. It can be seen that the FSS with narrower bandwidth will be less affected after HS-FSMSE is added. Hence, cascading HS-FSMSE layers to suppress spurious transmission of band pass FSS filters could be used for the applications which needs relatively narrow pass band.

It is worth pointing out that the unit cell size of the patch-grid FSS and the complementary loop FSS are different. The patch-grid FSS is sub-wavelength structure whose period is only 6.6 mm, and the complementary loop is a conventional resonant structure whose period is 10 mm. For two different FSSs, the unit cell size of the harmonic suppression FSMSE is the same as 3.3 mm. As a result, when the HS-FSMSE is integrated with the patch-grid FSS, one unit corresponds to 2 × 2 HS-FSMSEs. And for the resonant complementary loop FSS, one unit corresponds to 3 × 3 HS-FSMSEs, as shown in Figures 8 and 10, respectively. In this case, the different implementations show the transplantability of the HS-FSMSE.

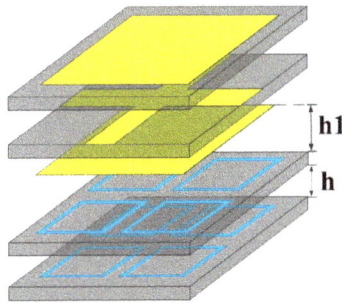

Figure 8. Multilayer patch-grid FSS integrating with HS-FSMSE (The yellow and blue parts represent metal, the grey represents substrates).

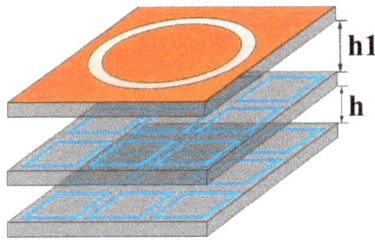

Figure 9. Complementary loop FSS integrating with HS-FSMSE (The orange and blue parts represent metal, the grey represents substrates).

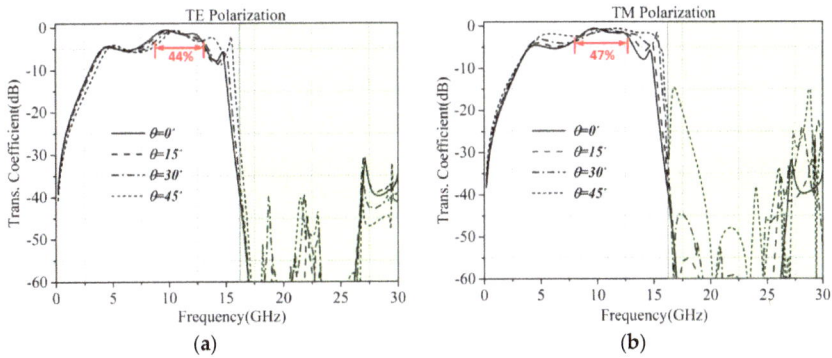

(a)   (b)

Figure 10. Transmission coefficient for oblique incidences with different angles on the multilayer patch-grid FSS with HF-FSMSE, (**a**) TE polarization and (**b**) TM polarization (The green shadow represents suppressed harmonic transmission bands).

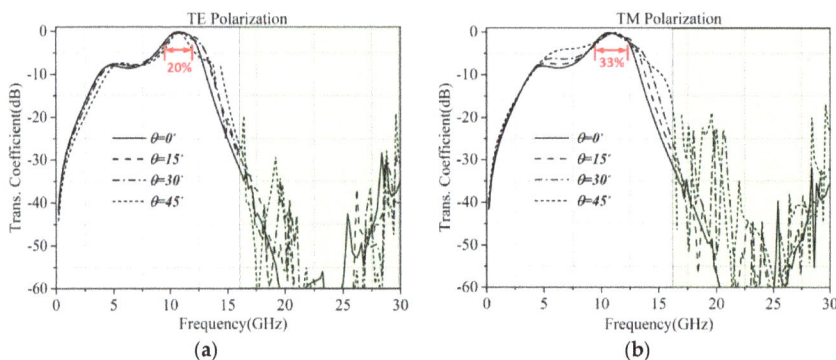

Figure 11. Transmission coefficient for oblique incidences with different angles on the complementary loop FSS with HF-FSMSE, (**a**) TE polarization and (**b**) TM polarization (The green shadow represents suppressed harmonic transmission bands).

3. Experimental Verification

In this section, a prototype of the patch-grid FSS integrating with HS-FSMSE has been fabricated, as shown in Figure 12. The fabricated prototype has five layers of metal structure printed on four substrates with dielectric constant $\varepsilon_r = 2.65$ and the loss tangent of 0.003. The layers are separated by air with plastic screws. The total size of the fabricated FSS is $235 \times 235 \times 7$ mm. The air spacing is 3 and 2, respectively. Since the dimension of the fabricated FSS is relatively small especially in the far-field environment which will make the diffraction becomes strong, we use the two-port method in the near-field region to measure the transmission feature of FSS. In the measurement, we measured the transmission and reflection characteristics of the two horns measurement system without FSMSE for calibration. The near field coupling effect can be eliminated. For frequency selective surfaces, the features do not vary with different kinds of electromagnetic wave. The plane wave setup in the simulation is mean to model the far-field environment. Since we have removed the near-field coupling effect, the measurement results could be equal to the theoretical model when the plane wave is incident on the FSS.

Experiments were performed by using three pairs of transmitting and receiving horn antennas that can cover different frequency ranges from 2 to 30 GHz. In these measurements, the broadband horns (2–18 GHz), K-band (18–26.5 GHz) horns and Ka-band (26.5–30 GHz) horns were used to measure the transmission coefficients of the fabricated FSMS. The system configuration for the measurement is shown in Figure 13, where Tx and Rx stand for the transmitting and receiving antennas, respectively, and RUT represents the fabricated FSS which was tested in the microwave chamber. The distance from Rx to FSS is 100 mm and from Tx to RUT is also 100 mm. It is worth noting that the center of RUT should be in direct line with that of Rx and Tx. The Anritsu MS46322A (E, Anritsu, Atsugi-shi, Kanagawa, Japan, 2015) 40 GHz vector network analyzer is used to measure the transmission characteristics. And we also measured the transmission and reflection characteristics of only two horns system without the FSS for calibration. Figure 14 shows the comparison of measured transmission coefficients between the original patch-grid FSS with and without HS-FSMSE. It can be seen that the designed HS-FSMSE effectively removes the undesired spurious transmission windows of the original patch-grid FSS. Even if the oblique incident angle is as large as 60 degree, the HS-FSMSE still works well. The measured frequency responses after calibration agree well with the simulated ones. The contribution of the proposed HS-FSMSE for the complementary resonant loop FSS is almost the same.

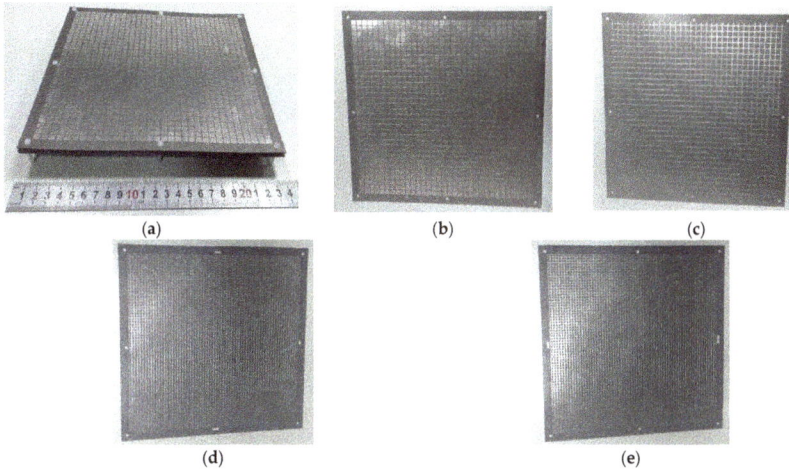

Figure 12. Geometry of the fabricated multilayer patch-grid FSS integrating with HS-FSMSE, (**a**) total view, (**b**) patch layer, (**c**) grid structure layer, (**d**,**e**) harmonic suppression FSMSE layers.

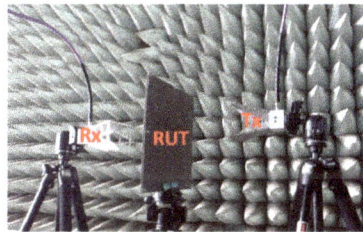

Figure 13. Measurement system setup for the FSS in the near field region.

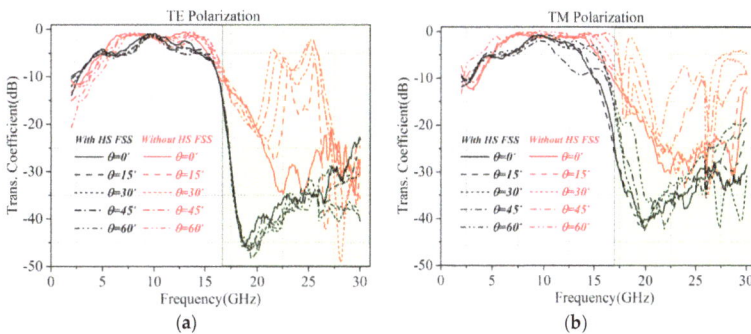

Figure 14. Comparison of the measured transmission coefficients between the non-resonant FSS with HS-FSMSE and without HS-FSMSE for oblique incidence angles for the (**a**) TE polarization and (**b**) TM polarization (The green shadow represents suppressed harmonic transmission bands).

4. Conclusions

A transplantable harmonic suppression FSMSE is designed and analyzed, which can be applied to arbitrary FSSs operating at the same frequency band. And the designed HS-FSMSE was integrated with two kinds of different FSSs: the sub-wavelength patch-grid FSS and the complementary resonant loop FSS. The simulation and measurement results show that the proposed HS-FSMSE can effectively remove the spurious transmission windows outside of the desired operation frequency range. Using the HS-FSMSE, the two FSSs operating at 10 GHz are free of high-order harmonic transmission bands up to 30 GHz without deteriorating the operating frequency of the pass band. The transplantable HS-FSMSE also shows good polarization and angle stabilities. The most essential cause of its flexibility lies in that the harmonic suppression FSMSE is designed independently from the original FSS. Additionally, the narrow band pass feature is taken in consideration when designing the harmonic suppression FSMSE. In other words, the design procedure of the original band pass FSS and harmonic suppression FSMSE are independent of one another. This makes the harmonic suppression FSMSE transplantable to different kinds of FSSs. What's more, the period of the harmonic suppression FSMSE is less than 0.2 wavelengths at band pass operating frequency (In this paper, the period of the harmonic suppression FSMSE is 0.11 wavelengths at band pass operating frequency). The small period size could be used for arranging most of the band pass FSS directly without cutting out both the harmonic suppression FSMSE and the original FSS.

Acknowledgments: This work is supported by National Natural Science Foundation of China under Contract No. 51477126, and supported by Technology Explorer and Innovation Research Project, and Fundamental Research Funds for the Central Universities.

Author Contributions: All authors contributed substantially to the reported work. Na Kou conceived the idea of transplantable high-order harmonic suppression frequency selective surfaces and performed the simulations and experiments; Haixia Liu provided the main instructions of experiments; Long Li provided the main instructions of this study and revised the paper.

Conflicts of Interest: The authors declare no conflict of interest.

References

1. Munk, B.A. *Frequency Selective Surfaces: Theory and Design*; Wiley-Interscience: Hoboken, NJ, USA, 2000.
2. Munk, B.A. *Frequency-Selective Surface and Grid Array*; Wiley: Hoboken, NJ, USA, 1995.
3. Costa, F.; Monorchio, A. A Frequency selective radome with wideband absorbing properties. *IEEE Trans. Antennas Propag.* **2012**, *60*, 2740–2747. [CrossRef]
4. Sievenpiper, D.; Zhang, L.; Broas, R.F.J.; Alexopolous, N.G.; Yablonovitch, E. High-impedance electromagnetic surfaces with a forbidden frequency band. *IEEE Trans. Microw. Theory Tech.* **1999**, *47*, 2059–2074. [CrossRef]
5. Lin, Y.; Wang, L.; Gao, J.; Lu, Y.; Jiang, S.; Zeng, W. Broadband working-waveband-tunable polarization converter based on anisotropic metasurface. *Appl. Phys. Express* **2017**, *10*, 032001. [CrossRef]
6. Chakravarty, S.; Mittra, R.; Williams, N.R. On the application of the microgenetic algorithm to the design of broad-band microwave absorbers comprising frequency-selective surfaces embedded in multilayered dielectric media. *IEEE Trans. Microw. Theory Tech.* **2001**, *49*, 1050–1059. [CrossRef]
7. Al-Joumayly, M.; Behdad, N. A Generalized method for synthesizing low-profile, band-pass frequency selective surfaces with non-resonant constituting elements. *IEEE Trans. Antennas Propag.* **2010**, *58*, 4033–4041. [CrossRef]
8. Abadi, S.M.A.M.H.; Li, M.; Behdad, N. Harmonic-suppressed miniaturized-element frequency selective surfaces with higher order bandpass responses. *IEEE Trans. Antennas Propag.* **2014**, *62*, 2562–2571. [CrossRef]
9. Pozar, D.M. *Microwave Engineering*, 3rd ed.; John Wiley & Sons: Hoboken, NJ, USA, 2004.
10. Parker, E.A.; Hamdy, S.M.A.; Langley, R.J. Arrays of concentric rings as frequency selective surfaces. *Electron. Lett.* **1981**, *17*, 880–881. [CrossRef]
11. Overview of Curve Fitting Models and Methods in LabVIEW. Available online: http://www.ni.com/white-paper/6954/en/ (accessed on 31 July 2009).

applied sciences

Article

MDPI

Electromagnetic Power Harvester Using Wide-Angle and Polarization-Insensitive Metasurfaces

Xuanming Zhang, Haixia Liu * and Long Li *

Key Laboratory of High Speed Circuit Design and EMC of Ministry of Education, School of Electronic Engineering, Xidian University, Xi'an 710071, China; zhd9988@163.com
* Correspondence: hxliu@xidian.edu.cn (H.L.); lilong@mail.xidian.edu.cn (L.L.);
  Tel.: +86-029-8820-4458 (H.L.); +86-029-8820-1157 (L.L.)

Received: 16 February 2018; Accepted: 22 March 2018; Published: 26 March 2018

Abstract: A new wide-angle and polarization-insensitive metasurface (MS) instead of traditional antenna is built as the primary ambient energy harvester in this paper. The MS is a two-dimensional energy harvesting array that is composed of subwavelength electrical small ring resonator that is working at 2.5 GHz (LTE/WiFi). In the case of different polarization and incidence angles, we demonstrate the metasurface can achieve high harvesting efficiency of 90%. The fabricated prototype of 9 × 9 MS energy harvesting array is measured, and the experimental results validate that the proposed MS has a good performance more than 80% of energy harvesting efficiency for arbitrary polarization and wide-angle incident waves. The good agreement of the simulation with the experiment results verifies the practicability and effectiveness of the proposed MS structure, which will provide a new source of supply in wireless sensor networks (WSN).

Keywords: metasurface (MS); energy harvesting; wide-angle; polarization-insensitive

1. Introduction

With the rise of the internet of things (IoT) technology, the micro wireless sensor networks (WSN) are being extensive development. In recent years, with the development of the ultra-low power consumption chip technology, the power consumption of wireless sensor node has already entered a microwatt magnitude range. At the same time, an increasing number of wireless network and mobile communication base stations and other radio transmitting equipment that is full of rich electromagnetic energy in the surrounding environment. It is also a research hotspot to reduce the demand of the battery by absorbing the ambient energy to power a sensor node automatically and adaptively in the electromagnetic environment.

Current researches show that there are about 12 million wireless communication base stations in the global in 2012, about 68 billion equipments using a radio frequency identification card (RFID) in 2013, and the world has about 40 billion WiFi consumer electronic equipments in 2014 [1]. Figure 1 shows that the ambient power spectrum distribution of the electromagnetic field measured in the campus environment of Xidian University, China. We can find that the ambient energy peaks appear in the existing wireless communication frequency bands. These micro electromagnetic energies that distribute in the environment are expected to be used again after harvesting efficiently.

Metamaterials are the artificial composite structures or composite materials with superior physical properties that some natural materials do not have. Metasurfaces (MS) are two-dimensional metamaterial structures, which have been attracting more attention, due to a wide range of potential applications for wireless power transfer and harvesting [2]. A perfect metamaterial absorber has been firstly proposed in 2008 by Landy et al. [3]. Different from the electromagnetic metamaterial absorber [4–6], the small split-ring resonators (SRR) were first used to collect energy in 2012 [7]. After that, some other SRRs [8–10], electric-inductive-capacitive (ELC) resonators [11], and ground-backed complementary split-ring (G-CSRR) resonator have also been proposed for electromagnetic energy harvesting [12,13]. The recently reported wideband ground-backed complementary split-ring (WG-CSRR) resonator structure can improve the bandwidth [14]. Each unit in the WG-CSRR structure requires four ports in order to collect energy at the same time.

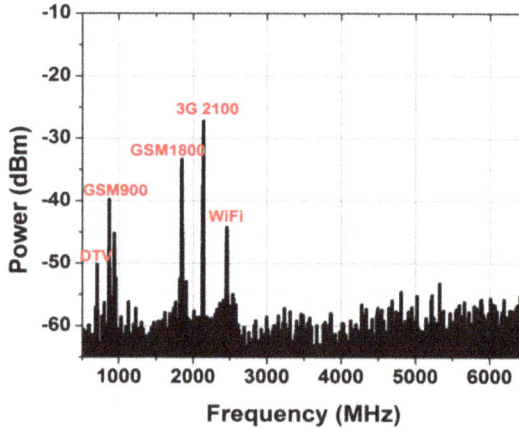

Figure 1. Power spectrum distribution measured in the campus of Xidian University, China.

This paper uses a new and simple electromagnetic MS as the ambient energy harvester in the wireless communication systems. The proposed MS is a two-dimensional energy harvesting array that is composed of subwavelength electrical small square-ring resonator unit. Because the incident angles and polarizations of environmental electromagnetic waves are random and unknown, the conventional energy harvesting antenna or polarization-related metamaterial harvester cannot maintain the high collection efficiency for arbitrary direction and polarization. No matter whether the incident wave is transverse electric (TE) or transverse magnetic (TM) polarization mode, the proposed MS unit or array can keep a high RF power harvesting efficiency. Due to the frequency distribution of electromagnetic energy in the environment has the narrow-band characteristics, this work uses only one harvesting port designed at the appropriate location for spatial filtering, which can simplify the back-end matching and rectifier circuit complexity, and realize the high efficiency of energy harvesting. The proposed MS is sub-wavelength, wide-angle, and polarization-insensitive, which is suitable for ambient power harvesting.

2. Design of the MS Energy Harvester

In the energy harvesting system, the rectenna is the most critical part [15–17]. It can be divided into energy harvester (e.g., antenna) and rectifier circuit. As the primary energy collector, it receives electromagnetic RF energy, which is converted to DC power through a rectifier circuit. The energy harvesting using MS should meet the following two conditions: (a) The electromagnetic wave energy can be captured by the MS structure as much as possible; (b) The captured energy into the MS structure can effectively be transferred into the rectifier circuits (matching load).

This paper builds a two-dimensional energy harvesting array that is composed of sub-wavelength square-ring MS, and each small unit can be equivalent to LC resonance circuit. The incident wave is stimulated to produce surface current on the surface of the MS structure. Similar to the antenna-based coaxial feed, we take a via hole in the unit with an appropriate place and then lead the surface current to the port that is connected to the resistive load [11]. The MS energy collector can capture electromagnetic wave energy and transfer to the load, rather than dissipate in the structure. The designed MS structure is shown in the Figure 2. The center working frequency is set to 2.5 GHz (LTE/WiFi). The upper layer is the square-ring array, and the middle is the low-loss dielectric layer and the bottom layer is the metallic ground. The energy harvesting ports are set at the corner of each ring.

Using ANSYS HFSS to establish the numerical simulation model, the boundary condition of the unit cell is set as periodic boundary conditions (PBC), which is combined with the Floquet ports to calculate the infinite array model. The port excitation of the incident wave power is set to 1 W. The dielectric substrate is F4B that has relative permittivity of 2.65 and a dielectric loss tangent of 0.001. The dimensions of the square-ring element are as follows: p = 15.7 mm, l = 15 mm, w = 0.8 mm, and t = 3 mm. Figure 3 shows that the calculated efficiency of the designed MS array for various load resistance at the operating frequency of 2.5 GHz. It can be seen that the highest efficiency of the operating frequency can be simultaneously obtained when the load resistance is 500 Ω. It should be noted that this matching load value is a local optimal solution. Other parameters, such as MS structure size, dielectric constant, and thickness of substrate, etc., will also affect the matching load value and the operating frequency. By changing these parameters, we can flexibly design the matching load value of rectifier circuit. This port is located in the diagonal direction of the square-ring element. Note that we can flexibly change the working frequency band, matching the load value by changing the value of each parameter of the MS structure.

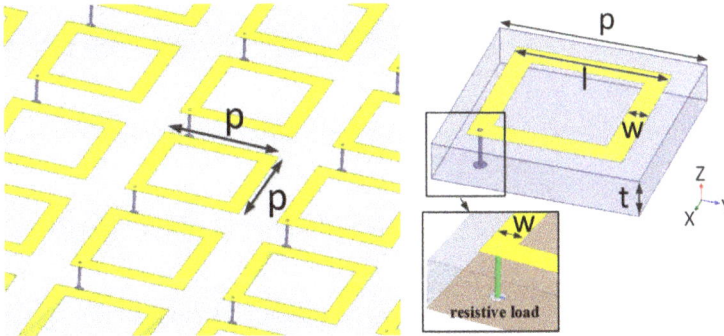

Figure 2. The square-ring resonator metasurface (MS) structure for ambient power harvesting.

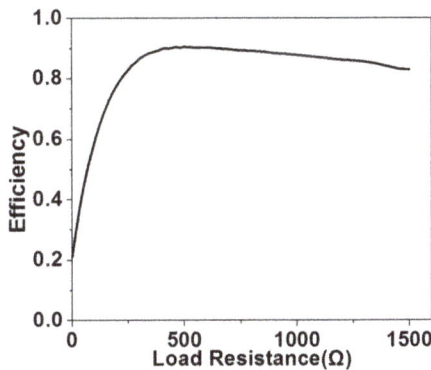

Figure 3. Calculated efficiency of the designed MS array for various values of load resistance at the operating frequency of 2.5 GHz.

When compared to the periodic dimension of the previous works, such as the SRR element, is about 1/4 operating wavelength at 5.8 GHz [7], and the G-CSRR element is about 1/3 operating wavelength at 5.55 GHz [13], and the WG-CSRR element is about 1/5 operating wavelength at 5.6 GHz [14], the proposed design in this paper is about 1/8 operating wavelength at 2.5 GHz, which is more miniaturization. Note that the proposed closed square-ring resonant ring rather than the split-ring form from existing literature is adopted in this paper. In spite of the angles and polarizations of the incident wave, two symmetrical annular currents in the same direction excited by an incident wave on the surface of the closed-ring can channel to the collection port. The open-ended surface currents distribution that was generated by various incident waves from the split-ring resonant lead to single-port collection with angle and polarization sensitive performance on a split-ring structure. So, the proposed square-ring MS structure can harvesting wide-angle, polarization-insensitive incident waves by optimizing the position of the harvesting point, and maintain high efficiency when compared to the structure described in the reference [6,10–12] only collects single-polarized incident waves. By changing the value of each parameter of the MS structure (L, W, P, ε_r, t), we can flexibly change the working frequency band, matching the load (R) value, to achieve the performance of effective energy harvesting. The square-ring MS structure is proposed to replace the traditional antenna as the front of the rectifier in the energy-harvesting system. It is worth pointing out that the proposed MS structure can effectively captures the ambient radiofrequency energy and then concentrated on the load resistance, which can be replaced by the rectifier circuit in the later work.

The power harvesting efficiency of the MS collector can be computed by

$$\eta = \frac{P_L}{P_r} \times 100\% \tag{1}$$

where P_L is the power collected by the load, and the P_r is the received total power incident on the cross-sectional area (A) of the whole MS structure, which can be computed using the surface integral of a Poynting vector on A.

The simulation results of the reflection coefficient S_{11} and power harvesting efficiency are shown in Figure 4. It can be seen that the impedance matching is good at the working frequency of 2.5 GHz. That is to say, the MS structure has the ability of capturing the electromagnetic wave energy. By calculating the receiving power in the lumped port, we can see that the MS harvester can collect the energy more than 90% from the incident power at 2.5 GHz. The captured electromagnetic energy by the square-ring MS can be transferred into the matching load through vias effectively. Figure 5 is the surface current distribution on the square ring and the back ground of the MS structure at 2.5 GHz. It can be seen that the surface current flows along the small ring, and the energy is eventually harvested to the port of the resistive load.

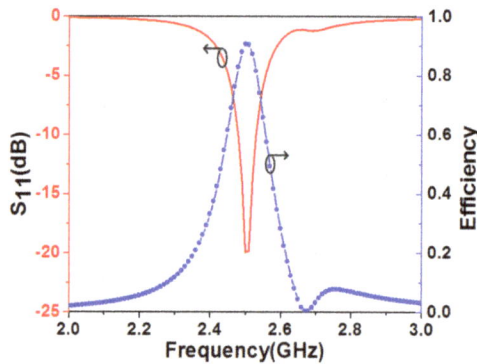

Figure 4. Reflection coefficient S_{11} and power harvesting efficiency using square-ring MS structure.

The incident angle and polarization stability of MS structure should be considered from the two cases of transverse electric (TE) and transverse magnetic (TM) oblique incidences. The TE-polarized oblique incidence refers to the change of electromagnetic wave incident angles to ensure the electric field vector is always parallel with the surface of MS structure, while the TM-polarized oblique incidence refers to the change of the electromagnetic wave incident angles to ensure that the magnetic field vector is always parallel with the surface of MS structure. Figure 6a shows the power harvesting efficiency characteristics at oblique incidence with TE polarization in the resonant bands. When the harvester tilted with angle θ = 0° and 15°, the maximum harvesting efficiency achieved was 90.5% at 2.5 GHz and 90.4% at 2.52 GHz. For angle θ = 30° and 45°, collection efficiency of 89.8% at 2.56 GHz and 88.4% at 2.62 GHz. The last tilted angle is θ = 60° and a collection efficiency of 85.2% is achieved at 2.67 GHz. For TM polarization, as shown in Figure 6b, the energy collection efficiency is 90% (when θ = 0° at 2.5 GHz), and the maximum efficiency is 91.4% (when θ = 45° at 2.62 GHz). When the oblique incidence θ are 15°, 30°, and 60°, the harvesting efficiency are 91.1% at 2.52 GHz, 91% at 2.56 GHz, and 91.2% at 2.67 GHz. The rate of change of efficiency is very small. As the angle of incidence waves increases, the working frequency of the maximum efficiency obtained by the MS will slight shift. The MS have a certain bandwidth. No matter TE or TM oblique incidence from 0° to 60°, the energy harvesting efficiency using the square-ring MS collector is always stable and the collection efficiency can reach 90%. Though it has a certain frequency-shift in the incidence of the large angles, it can be adjusted through backend self-adaptive tuning circuits.

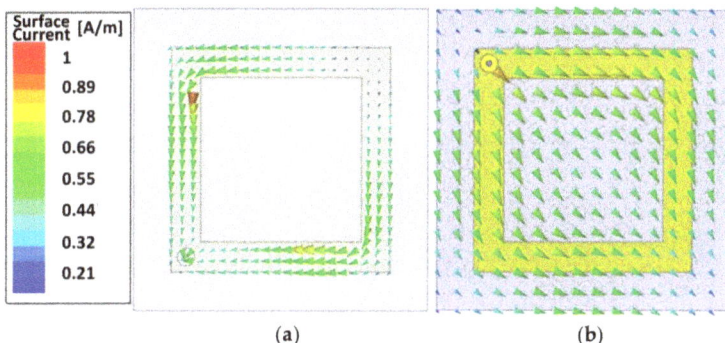

Figure 5. Surface current distribution on the (**a**) square ring and (**b**) back ground of the MS structure at the resonant frequency (2.5 GHz).

The surface current analysis can be used to gain insight into the polarization and the angle stability of the MS. The resonant behavior can be described by using the equivalent inductance (L) and the equivalent capacitance (C) of the MS structure. The small square-ring of each unit and gap between units can be equivalent to LC resonance circuit. The surface currents induced on the front metallic layer operating at work frequency of 2.5 GHz are oscillated along the square ring of each element and transferred to the load through vias. The surface currents in TE and TM polarization of the incident wave at the resonant frequency (2.5 GHz) and non-resonant frequency (2.2 GHz) for the incidence angle 0° are shown in Figure 7, respectively. It can be seen that the maximum intensity of the surface induced current to the load is generated on the metallic ring at operating frequency of 2.5 GHz, and the maximum energy harvesting efficiency can be obtained. When offsetting the operating frequency, the intensity of the surface induced current on the metallic ring is reduced, which causes the load absorption decreased. Figure 8 shows the surface current distribution at the working frequency for different incidence angles in TE polarization. Maximum energy harvesting efficiency with θ = 30° and 60° of incident wave at 2.56 GHz and 2.67 GHz, respectively. Because the square-ring MS is centrosymmetric, no matter whether the incident wave is TE or TM polarization with large angle range, it can excite surface currents along the metal ring and translate into load by vias for efficient energy harvesting.

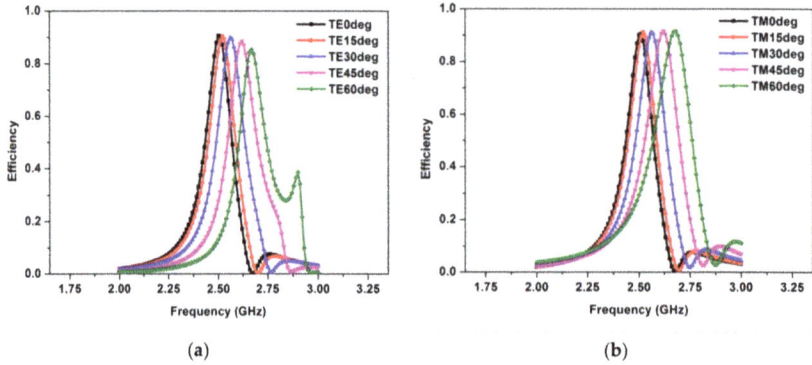

Figure 6. Power harvesting efficiency of (**a**) transverse electric (TE)-polarized oblique incidence and (**b**) transverse magnetic (TM)-polarized oblique incidence.

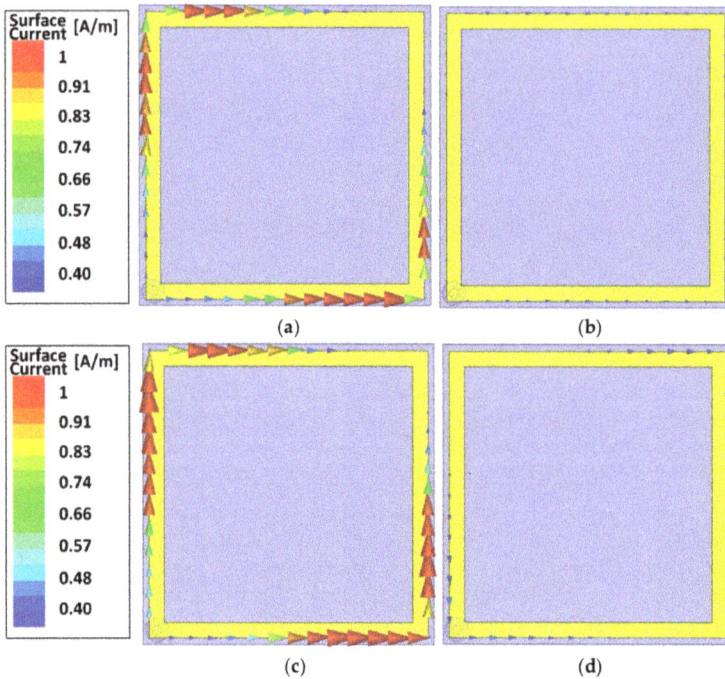

Figure 7. Simulated surface current distribution on the square-ring at (**a**) the resonant frequency of 2.5 GHz with TE polarization; (**b**) the non-resonant frequency of 2.2 GHz with TE polarization; (**c**) 2.5 GHz with TM polarization; and (**d**) 2.2 GHz with TM polarization.

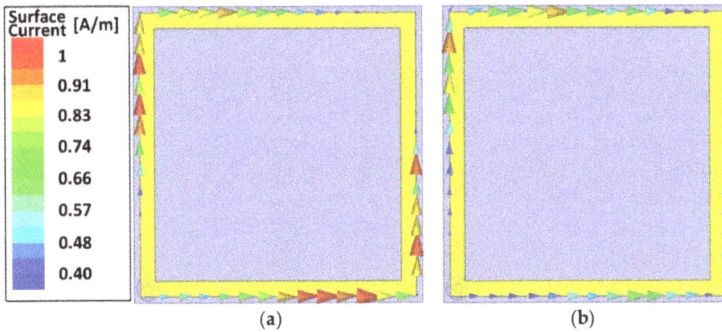

Figure 8. Simulated surface current distribution on the square-ring with different incidence angles (a) θ = 30°; (b) θ = 60°.

3. Experimental Verification

In this section, a 9 × 9 MS energy harvesting array has been fabricated to verify the performance, as shown in Figure 9. The experimental scheme is illustrated in Figure 10. The MS energy harvester placed in the far field of the horn antenna that excited by a signal generator at the power level of 14 dBm, to ensure that the electromagnetic wave incident to the MS structure is a plane wave excitation [11]. The power of the load resistance on the MS structure is tested by using a spectrum analyzer. Because each cell of the 9 × 9 MS array has a harvesting port to collect energy at the same time, the power measurement for all of the ports is almost impossible. The central unit can represent other units in a large array without edge effects. Therefore, a good estimation of the power harvesting can be measured. Based on the analysis and discussion above, we only test energy harvesting efficiency of the center element, and the overall efficiency of array can be calculated by multiplying the efficiency of the center element by the number of array elements [13].

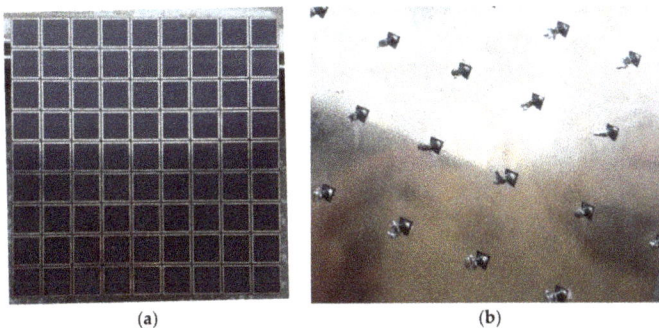

Figure 9. Photograph of the fabricated MS structure array on (a) front view (b) back view.

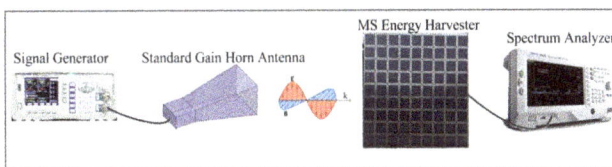

Figure 10. Experimental setup for measuring the power harvesting efficiency of MS structure.

The energy collection efficiency is given by:

$$\eta = \frac{P_{RF}}{P_{MS}} \times 100\% \tag{2}$$

where P_{RF} is the total time-average power harvested by the port of central element (measured by the spectrum analyzer). P_{MS} is the total time-average power incidence on the center cell of the MS by the standard gain horn antenna.

Figure 11 shows the measured energy collection efficiency of the fabricated 9 × 9 MS energy harvesting array in different angles of TE and TM oblique incidence from the same distance. It can be seen that the collected energy efficiency is more than 80% in the TE normal incidence at 2.58 GHz. When the incident angle is changed to θ = 60°, the energy collection efficiency in this case is still 81%, but the operational frequency will shift to 2.7 GHz. In the TM polarization, we can see that the maximum collection efficiency is 85.1% and the resonant frequency is 2.57 GHz when the incident angle θ is 0°. When the incident angle changes to θ = 60°, the resonant frequency of maximum efficiency shifts to 2.69 GHz, and the collection efficiency is 82.6%. It is verified that the proposed MS structure has a good performance on the incident electromagnetic wave with arbitrary polarization. The energy harvesting efficiency and the operational frequency that was obtained from the measurement and the numerical analysis are different by reason of typical fabrication tolerances and measurement error. Moreover, the fabricated array is a 9 × 9 array when compared to infinite periodic array in the simulation. So, the measurement efficiency is lower than the simulation and the operating frequency have a slight offset, but the energy collection efficiency remains above 80%. The frequency shift can be adjusted using self-adaptive tuning circuits. The good agreement of the simulation with the experiment results verifies the practicability and the effectiveness of the proposed MS structure.

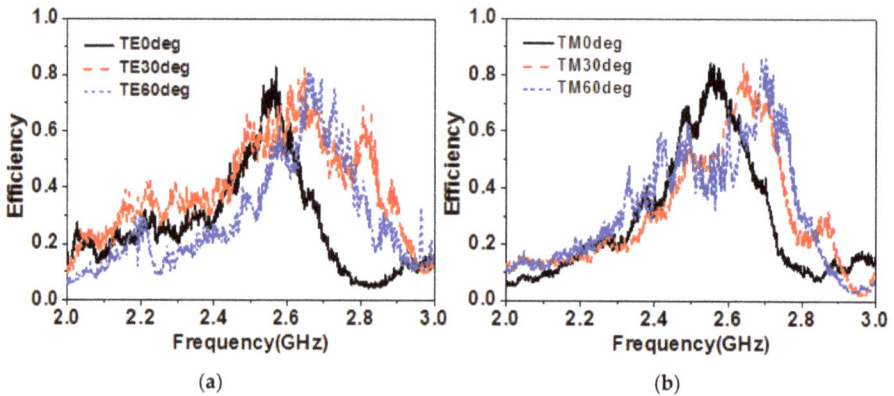

Figure 11. Measured energy harvesting efficiency of the fabricated 9 × 9 MS energy harvesting array for different polarization and incident angles, (**a**) TE-polarized oblique incidence and (**b**) TM-polarized oblique incidence.

4. Conclusions

In this paper, a new polarization-insensitive MS was designed, fabricated, and experimentally demonstrated to absorb the electromagnetic power in wide incidence angle. The designed MS is an electrical small square-ring resonator whose center operating frequency is 2.5 GHz (LTE/WiFi). No matter whether the incident wave is TE or TM polarization, the proposed MS can keep high RF power harvesting efficiency. The good performance and accuracy of the proposed MS design are verified by using simple and effective simulations and measurements. The proposed square-ring MS harvester has many advantages, such as miniaturization, polarization insensitivity, wide angle stability, high efficiency, and easy to conformal, which are suitable for ambient power harvesting.

Acknowledgments: This work is supported by National Key R&D Program of China, and is supported by National Natural Science Foundation of China under Contract No. 51477126, and supported by Technology Explorer and Innovation Research Project.

Author Contributions: All authors contributed substantially to the reported work. Long Li conceived the idea of metasurface for ambient energy harvesting. Xuanming Zhang designed the subwavelength electrical small ring resonator and performed the simulations and experiments; Haixia Liu provided the main instructions of experiments; Long Li provided the main instructions of this study and revised the paper.

Conflicts of Interest: The authors declare no conflict of interest.

References

1. Hemour, S.; Wu, K. Radio-frequency rectifier for electromagnetic energy harvesting: Development path and future outlook. *Proc. IEEE* **2014**, *102*, 1667–1691. [CrossRef]
2. Li, L.; Liu, H.; Zhang, H.; Xue, W. Efficient wireless power transfer system integrating with metasurface for biological applications. *IEEE Trans. Ind. Electron.* **2018**, *65*, 3230–3239. [CrossRef]
3. Landy, N.I.; Sajuyigbe, S.; Mock, J.J. A perfect metamaterial absorber. *Phys. Rev. Lett.* **2008**, *100*, 207402. [CrossRef] [PubMed]
4. Ding, F.; Cui, Y.; Ge, X.; Jin, Y.; He, S. Ultra-broadband microwave metamaterial absorber. *Appl. Phys. Lett.* **2012**, *100*, 103506. [CrossRef]
5. Luukkonen, O.; Costa, F.; Monorchio, A.; Tretyakov, S.A. A Thin Electromagnetic Absorber for Wide Incidence Angles and Both Polarizations. *IEEE Trans. Antennas Propag.* **2009**, *57*, 3119–3125. [CrossRef]
6. Li, L.; Yang, Y.; Liang, C. A wide-angle polarization-insensitive ultra-thin metamaterial absorber with three resonant modes. *J. Appl. Phys.* **2011**, *110*, 063702. [CrossRef]
7. Ramahi, O.M.; Almoneef, T.S.; AlShareef, M.; Boybay, M.S. Metamaterial particles for electromagnetic energy harvesting. *Appl. Phys. Lett.* **2012**, *101*, 173903. [CrossRef]
8. Hawkes, A.M.; Katko, A.R.; Cummer, S.A. A microwave metamaterial with integrated power harvesting functionality. *Appl. Phys. Lett.* **2013**, *103*, 163901. [CrossRef]
9. Alshareef, M.R.; Ramahi, O.M. Electrically small resonators for energy harvesting in the infrared regime. *J. Appl. Phys.* **2013**, *114*, 223101. [CrossRef]
10. Shang, S.; Yang, S.; Liu, J. Metamaterial electromagnetic energy harvester with high selective harvesting for left- and right-handed circularly polarized waves. *J. Appl. Phys.* **2016**, *120*, 045106. [CrossRef]
11. Almoneef, T.; Ramahi, O.M. Metamaterial electromagnetic energy harvester with near unity efficiency. *Appl. Phys. Lett.* **2015**, *106*, 153902. [CrossRef]
12. Alavikia, B.; Almoneef, T.S.; Ramahi, O.M. Electromagnetic energy harvesting using complementary split-ring resonators. *Appl. Phys. Lett.* **2014**, *104*, 163903. [CrossRef]
13. Alavikia, B.; Almoneef, T.S.; Ramahi, O.M. Complementary split ring resonator arrays for electromagnetic energy harvesting. *Appl. Phys. Lett.* **2015**, *107*, 033902. [CrossRef]
14. Alavikia, B.; Almoneef, T.S.; Ramahi, O.M. Wideband resonator arrays for electromagnetic energy harvesting and wireless power transfer. *Appl. Phys. Lett.* **2015**, *107*, 243902. [CrossRef]
15. Chen, Z.; Guo, B.; Yang, Y.; Cheng, C. Metamaterials-based enhanced energy harvesting: A review. *Phys. B Condens. Matter* **2014**, *438*, 1–8. [CrossRef]
16. Olgun, U.; Chen, C.C.; Volakis, J.L. Design of an efficient ambient WiFi energy harvesting system. *IET Microw. Antennas Propag.* **2012**, *6*, 1200–1206. [CrossRef]
17. Chin, C.H.K.; Xue, Q.; Chan, C.H. Design of a 5.8-GHz rectenna incorporating a new patch antenna. *IEEE Antennas Wirel. Propag. Lett.* **2005**, *4*, 175–178. [CrossRef]

applied
sciences

MDPI

Article

Minkowski Island and Crossbar Fractal Microstrip Antennas for Broadband Applications

Roman Kubacki *, Mirosław Czyżewski and Dariusz Laskowski

Faculty of Electronics, Military University of Technology, 00-809 Warsaw, Poland;
miroslaw.czyzewski@wat.edu.pl (M.C.); dariusz.laskowski@wat.edu.pl (D.L.)
* Correspondence: roman.kubacki@wat.edu.pl

Received: 4 February 2018; Accepted: 24 February 2018; Published: 27 February 2018

Abstract: The paper presents microstrip patch antennas, which are based on the fractal antenna concept, and use planar periodic geometries, providing improved characteristics. The properties of the fractal structure were used in a single-fractal layer design as well as in a design, which employs fractals on both the upper and bottom layers of the antenna. The final structure, i.e., a double-fractal layer antenna has been optimized to enhance bandwidth and gain of the microstrip antenna. The proposed geometry significantly improved antenna performance. The antenna could support an ultra-wide bandwidth ranging from 4.1 to 19.4 GHz, demonstrating higher gain with an average value of 6 dBi over the frequency range, and a radiation capability directed in the horizontal plane of the antenna.

Keywords: metasurfaces; microstrip antenna; fractal antenna

1. Introduction

Antennas are a key element in wireless communication systems and one of the important problems in antenna design is to provide wideband characteristics. Telecommunication systems, especially those intended for military applications, should be equipped with an antenna capable of supporting a wideband or even ultra-wideband (UWB) frequency range. In addition, for some systems, such as cognitive radio, the antenna should allow for dynamically and autonomously emitting and/or receiving signals in the desired frequency ranges [1]. The efficiency of a radio broadcasting system is directly related to the characteristics of its antenna. Like other designs before them, microstrip antennas revolutionized antenna technology. Attractive properties of microstrip patch antennas—low profile, light weight, compact and conformable structure, and easy fabrication, make them good candidates for wireless systems. However, microstrip patch antennas have a narrow bandwidth, which is a significant drawback that limits their usefulness.

The conventional microstrip antenna consists of a metal strip conductor (patch) as an upper layer of a grounded substrate. The performance of microstrip antenna, in terms of bandwidth and radiation characteristics, is determined by its shape, dimensions and the dielectric constant of the substrate.

With the development of new materials called left-handed metamaterials (LHM), it is possible to create antenna designs with new properties. The microstrip antennas based on composite LHM should be characterized by simultaneously negative permittivity and permeability. Such materials exhibit counter-intuitive electromagnetic phenomena, such as reversed refraction. As a result, they can be referred to as negative refractive index (NRI) materials, since the wave vector forms a left-handed triplet.

Structures based on split ring resonators (SRR), with concentric rings separated by gaps in the ground plane, can yield to a material with simultaneously negative permittivity and permeability [2]. Such a metamaterial-based antenna can be smaller despite having improved properties. However, the antennas based on SRR and wire strips cannot achieve a sufficiently wide frequency range. This is why many different attempts have been proposed to widen the bandwidth of microstrip antenna using left-handed metamaterial structures. Among these, we can find a broadband planar antenna composed of unit cells and a dipole element having metamaterial property [3]. This configuration achieved a bandwidth of 1.2 GHz.

Planar left-handed metamaterial structures were proposed by Matsunaga et al. [4]. The presented structures consisted of 2D periodic arrays of unit cells on both the upper and lower layers. On the upper layer, 45-degree rotated square metal cells were placed, while, on the lower layer, four isosceles triangles were laid out and surrounded by a square frame. This configuration allowed for obtaining a frequency range, which is several times wider than one achieved by the same patch antenna without the periodic pattern.

Taking into account the fact that the main motivation of this work was to design an antenna with the widest possible bandwidth, the structure of the microstrip antenna was motivated by the advantages of using fractals in antenna engineering. The proposed antenna geometries were based on crossbar and Minkowski fractals, as an increased number of resonances over a given or fixed frequency band was reported for such antennas [5].

Fractal antennas are designs inspired by structures found in nature, and take up highly convoluted shapes. Fractal shapes applied in antenna designing can have various shapes and forms [3,6]. One can mention the following mathematically described shapes of fractals: Sierpinski carpet and gasket, Minkowski gasket, Koch snowflake, Mandelbrot set, crossbar, etc. These are the deterministic fractals, which can be used to create a scaled-down design with a recursive algorithm. The philosophy of fractal structure is that it can be subdivided into parts, with each part being a copy of the entire assembly, but reduced in size. One of the main benefits of the fractal structure is the increased electrical length of the antenna when compared to a conventional microstrip antenna. This property allows for the following advantages: it reduces the overall size of the antenna and produces multiple resonant bands. This benefit can be obtained due to the fractal geometry, as fractals have no characteristic size and could be treated as the composition of many copies of themselves at different scales. Best [5] pointed out that a fractal-shaped antenna demonstrates resonance compression and multi-band behavior. However, as it is underlined in [7], the fractal antenna with etched multiple slots suffers from the reduced overall gain, as a major part of the conducting strip is etched out.

Many types of fractal based microstrip antennas are presented in the literature, and their promising electrical properties were already revealed, but fractal structures were introduced mainly to the upper layers (patches), while the conducting ground planes were kept as monolithic strips.

Mosallaei et al. [8] and Yao et al. [9] presented microstrip antennas with periodic structures on the bottom layers. The bottom surface was constructed from periodic structures, based on square patches, which allowed for enhancing bandwidth despite the miniaturization of antennas.

Only a few papers discussed the electrical properties of microstrip antenna, where both the upper and the lower layers have periodic structures, as opposed to monolithic structures. Li-Wei et al. presented an antenna having upper and lower layers with periodic gaps designed in the form of isolated micro-triangles, while on the bottom ground plane has simple, periodically distributed cross strip-line gaps [10]. They obtained promising electrical properties—the working frequency bandwidth spanning from 200 MHz to 3 GHz.

In our previous work, we presented a microstrip antenna with both layers designed using periodic geometry [11]. The upper and lower layers of the antenna had fractal forms, but each fractal layer had a different shape. The Minkowski and crossbar fractals were taken into consideration.

In this work, the investigation of the electrical properties of fractal antennas with the following geometries was presented:

-   an antenna, in which the upper layer has fractal geometry while the lower layer has classical monolithic structure,
-   an antenna, in which the upper layer is a monolithic patch but the ground is a fractal,

- an antenna, in which both layers (upper and lower) have fractal structures, but with the same type of fractal,
- an antenna, in which both layers have fractal structures, each with a different type of fractal.

Taking into account the above demands, three kinds of antennas with fractals were investigated, i.e., crossbar fractal antennas, a Minkowski fractal antenna, and a crossbar and Minkowski fractals antenna, to recognize the impact of the type of fractal geometry on antenna properties. This also could be used to illustrate the benefits of fractal structures on the lower or upper layers and shows the level of proportional participation of each layer in the total electric properties of fractal antennas.

In all cases, the entire structure of the proposed antennas measures 27.6×31.8 mm^2. The antennas were designed with the RT/Duroid®5880 (Rogers Corporation, Killingly, CT, USA) low-loss substrate (tanδ = 0.0009 at 10 GHz) with dielectric constant of 2.2 and thickness of 0.79 mm.

2. The Conventional Reference Antenna

A rectangular microstrip patch antenna was designed to serve as a reference. The patch is 12×16 mm^2, and the feeding line is 2.46×8 mm^2 (Figure 1a). Note the asymmetrical geometry of the feeding line. Such configuration was mainly intended for the antenna with double-fractal layers presented in Figure 6. Nevertheless, all antennas analyzed below have the same asymmetrical geometry to ensure the same feeding condition. The simulated and measured values of reflection coefficient (S11 [dB]) of the conventional reference antenna are presented in Figure 1b.

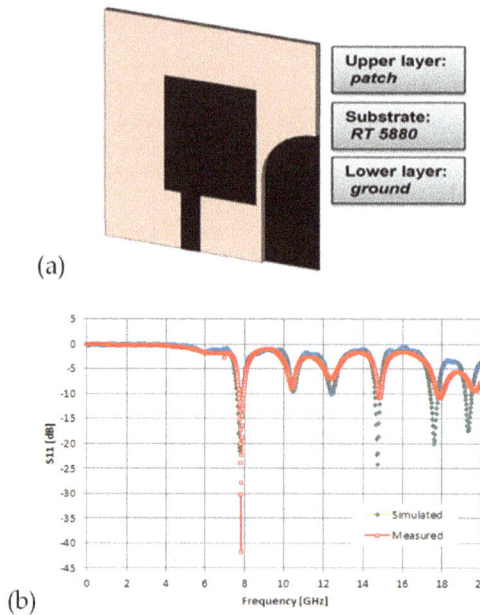

Figure 1. (**a**) Geometry of conventional microstrip antenna; (**b**) simulated and measured S11 of the conventional microstrip antenna.

The S11 drawing of conventional microstrip patch antenna (Figure 1a) shows a very limited bandwidth of 225 MHz (Figure 1b).

3. The Crossbar Fractal Geometry

The microstrip antennas based on crossbar fractals offer the enhanced frequency band but also the improved radiation antenna pattern. El-Hameed et al. presented such fractal antenna, receiving the reflection coefficient better than −10 dB in the range of 3.1–11.5 GHz [12]. Particular applications for such a fractal with different three-stages (configurations) were discussed in [12–14]. The advantage of a crossbar fractal is that the number of stages can be adjusted to obtain the required multi band characteristics.

In this work, stage-2 of crossbar fractal tree structure, diagonally lying on the square unit cell of 4 × 4 mm² has been taken into consideration—Figure 2a.

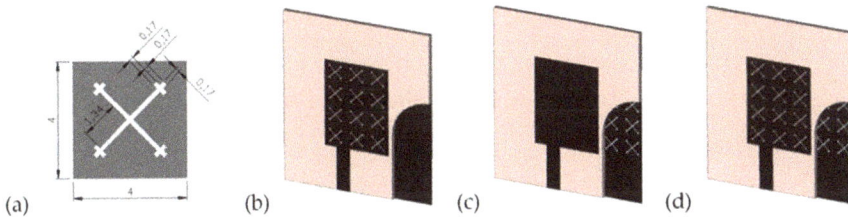

(a)   (b)   (c)   (d)

Figure 2. (a) the crossbar fractal based elementary unit cell of the antenna (dimensions are in mm); (b–d) geometry of the three configurations (mentioned in the text) of microstrip antennas with crossbar fractals.

The following antenna configurations have been analyzed:

(a) Configuration #1. The upper layer is based on a crossbar fractal structure and the lower layer is a solid conductive strip as in a conventional antenna. The upper patch consists of the 4 × 3 square unit cells with an edge of 4 mm in a periodic arrangement.

(b) Configuration #2. The patch is a solid strip, but the ground layer is composed of 7 × 6 square unit cells with the crossbar fractal.

(c) Configuration #3. Upper and lower layers are based on a crossbar fractal.

The above configurations of antennas have been presented in Figure 2b–d.

The reflection coefficients (S11 [dB]) for the above-mentioned configurations of antennas with crossbar fractal where depicted in Figure 3.

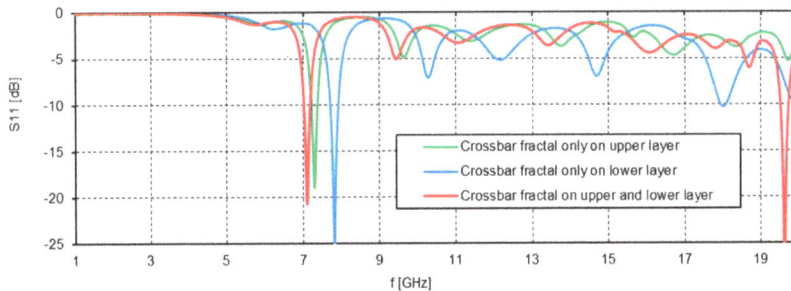

Figure 3. Simulated values of S11 for the antennas with crossbar fractals, for the three configurations.

Reflection coefficients for analyzed antennas indicate the narrowband behavior for all configurations. Such property is practically similar to the conventional microstrip antennas. The only small displacements in frequency can be observed, depending on the proportional participation of the metal being etched out. Figure 3 shows that the crossbar fractal cannot make it possible to broaden the frequency band of the antennas. Nevertheless, one property of this fractal should be underlined—the deepness of the S11. Curves presented in Figure 3 have significantly smaller depth compared to the conventional antenna (Figure 1b). Such crossbar fractal smoothing property will be used with the final version of an antenna with double different fractal layers.

4. The Minkowski Island Fractal Geometry

Minkowski gasket is one of the earliest structures introduced into the antenna technique and is the most promising structure. Dhar et al. presented Minkowski fractals based microstrip antenna allowing for covering five different wireless standards [7]. Geometrical structure of presented antenna was loaded with a nearly square shaped dielectric resonator allowing for obtaining constant gain at all discrete frequencies. Moraes et al. [15] presented fractal antenna structure, based on Minkowski squares, operated at four resonant frequencies for WiMAX systems while Shafe et al. [16] described Minkowski fractal structure at three frequencies for WiMAX, WLAN and HiperLAN applications. Singh et al. showed that fractals are able to enlarge the frequency antenna operation if their structures are self-similar having the same basic appearance at every scale [17]. Dhar et al. showed the geometry of Minkowski fractal allowing for obtaining the wide bandwidth [3].

The iterative-generation of Minkowski fractals was discussed in [5,18,19]. Comissio [20] and Best [21] noted that resonant frequencies can be modified by adjusting the level of fractal iteration. In this investigation, the geometry of unit cell with the first iteration of Minkowski island, as depicted in Figure 4a, will be considered. It should be emphasized that the antenna behavior with increasing fractal iterations yields to worse electrical properties, resulting in decreasing broadbandedness of the antenna.

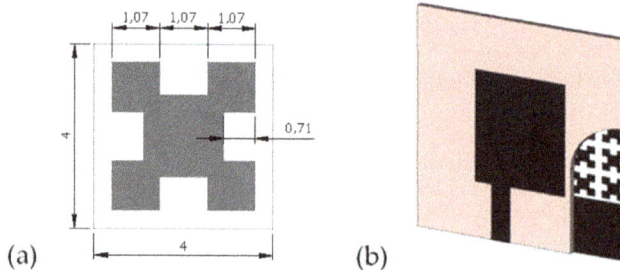

Figure 4. (**a**) The Minkowski island fractal based elementary unit cell of the antenna (dimensions are in mm); (**b**) geometry of the antenna with Minkowski fractal on the lower layer.

The Minkowski fractal based antenna was analyzed only for one configuration—the patch is a solid strip while the ground layer is composed of 7 × 6 square unit cells with Minkowski fractals, separated by slot gaps and placed in a periodic arrangement. This configuration of an antenna has been presented in Figure 4b.

In this work, the configuration when the upper layer is based on Minkowski fractal was not analyzed because of the problem with feeding. With the dimensions of Minkowski fractal (Figure 4a), the feeding line for the upper layer would have not the impedance of 50 Ω. Singh et al. [10] adapted the feeding technique to Minkowski fractal on the upper layer, but, in their solution, Minkowski fractal has modified geometry. Such modification results in different antenna geometry than that analyzed in this paper.

The reflection coefficients (S11 [dB]) for the above-mentioned configuration of antenna with Minkowski island fractal was depicted in Figure 5.

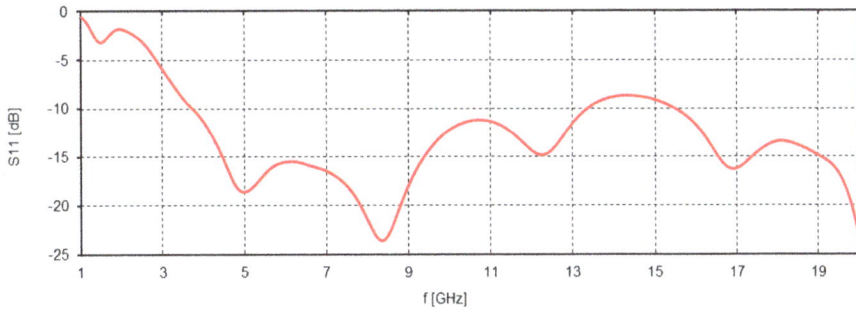

Figure 5. Simulated values of S11 for the antenna with ground based on Minkowski fractal.

As shown in Figure 5, the plot of S11 demonstrates good reflection properties of the antenna in the frequencies from 4 to 13.5 GHz and above 15.7 GHz. However, the band from 13.5 to 15.7 GHz in unacceptable due reflection properties being too high. This disadvantage will be reduced or mitigated in the final version of the antenna, with the patch based on a crossbar fractal.

5. The Double-Fractal Layers Geometry Antenna

Taking into consideration the above investigation, the antenna with double fractal layers has been designed. The upper layer is based on a crossbar fractal, as presented in Figure 2a, while the ground has a Minkowski island fractal as is depicted in Figure 4a. The upper layer with the feeding line has an asymmetrical geometry. This feeding configuration enables an undisturbed flow of current from the feeding line to the patch. The final version of the antenna is presented in Figure 6. Double-fractal geometry was combined to obtain the optimal configuration of the antenna, the best S11 performance, and thus the largest bandwidth.

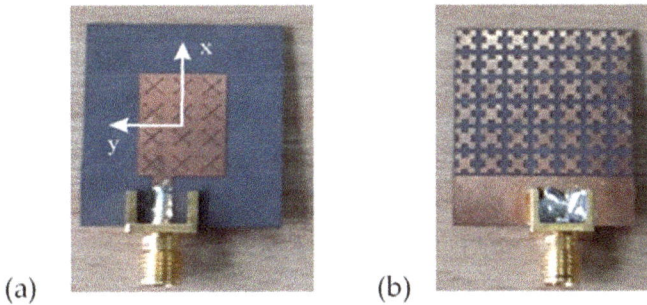

Figure 6. Picture of the fabricated antenna. Picture of the fabricated antenna. (**a**) One side of the antenna (**b**) another side of the antenna.

The reflection properties of the antenna have been presented in Figure 7.

Figure 7 demonstrates the widened bandwidth of the final version of the antenna, where the measured frequency range is even wider than the simulated one. It has achieved, at the level of −10 dB, a bandwidth ranging from 4.1 to 19.4 GHz, for a total bandwidth of 15.3 GHz. The figures show good correlation between measured and simulated values, and the very wide bandwidth of the proposed configuration.

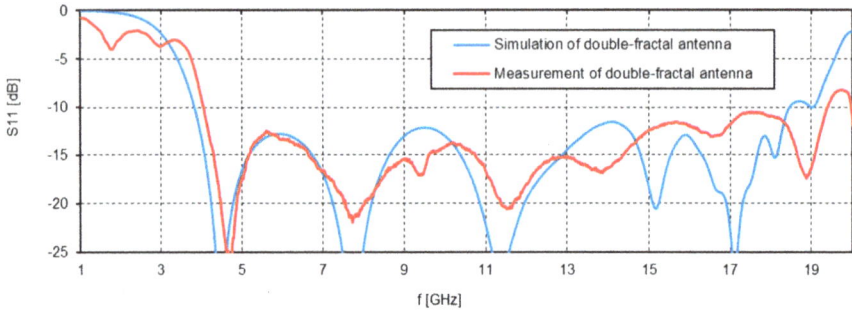

Figure 7. Results of simulated and measured values of S11 for the double-fractal antenna.

The gain (G) of the proposed antenna was measured over the working bandwidth, from 4 to 18 GHz. The antenna achieved an average measured gain of 6 dBi and a maximum value of 10.9 dBi at 15.2 GHz. For lower frequencies, the value of gain falls, e.g., G = 5.9 dBi at 10 GHz, G = 4.5 dBi at 6 GHz. The efficiency improves with the growth of frequency. The following values of efficiency were obtained: of 86.4% at 7.96 GHz, of 88.3% at 9.22 GHz and of 89.7% at 11.48 GHz.

The radiation characteristics for proposed antenna were presented in Figure 8. Radiation patterns were measured in frequencies of 7.96, 9.22 and 11.48 GHz in the characteristic planes—"X-Y" and "X-Z". These are the exampled frequencies for the frequency range of significantly low reflection property (7.96 and 11.48 GHz) as well as for the rather worse reflection coefficient (9.22 GHz). It can be seen that proposed geometry of the antenna resulted in a radiation spread of 270°. However, the shapes of radiation characteristics change in function of frequency. Future efforts should focus on obtaining similar shapes of radiation patterns in function of frequency to improve the efficiency of radio link.

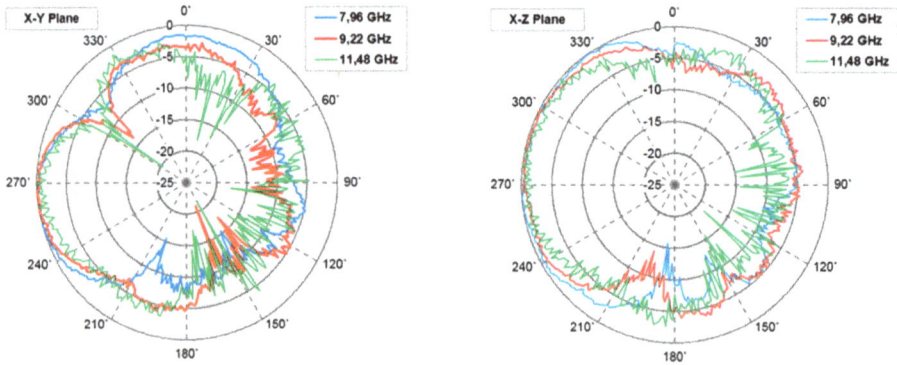

Figure 8. Results of measured radiation patterns of the antenna.

6. Conclusions

In the paper, ultra-wideband antennas based on the planar periodic fractal geometry concepts were presented.

The electrical properties of antennas with one layer based on fractals were analyzed to recognize the influence of the fractal geometry on the broadbandedness of the antenna. The crossbar and Minkowski island fractals were taken into consideration.

Appl. Sci. **2018**, *8*, 334

The final antenna, based on double fractal layers, achieved the frequency range from 4.1 to 19.4 GHz, realizing a bandwidth of 15.3 GHz. Therefore, the bandwidth of manufactured antenna is 68 times larger than the bandwidth of a conventional microstrip antenna.

The antenna radiates at a 270° angle. The combination of the planar concept of the fractal shapes seems to be a good method to enhance the characteristics of the microstrip patch antenna. This antenna can easily be used in a communication system, where such enhanced features are required.

Acknowledgments: This work was supported by the National Centre for Research and Development (NCBR) of Poland under project No. DOB-1-1/1/PS/2014.

Author Contributions: All authors contributed substantially to the reported work. Roman Kubacki was the originator of the idea of this study and prepared the manuscript; Mirosław Czyżewski and Dariusz Laskowski provided the numerical calculations and performed the experiments.

Conflicts of Interest: The authors declare no conflict of interest.

References

1. Swedheetha, C.; Suganya, M.; Gunapandian, P.; Manimegalai, B. Minkowski fractal based antenna for Cognitive Radio. In Proceedings of the 2014 IEEE International Microwave and RF Conference (IMaRC), Bangalore, India, 15–17 December 2014; pp. 166–169.
2. Lee, Y.; Hao, Y. Characterization of microstrip patch antennas on metamaterial substrates loaded with complementary split-ring resonators. *Microw. Opt. Technol. Lett.* **2008**, *50*, 2131–2135. [CrossRef]
3. Dhar, S.; Ghatak, R.; Gupta, B.; Poddar, D.R. A Wideband Minkowski Fractal Dielectric Resonator Antenna. *IEEE Trans. Antennas Propag.* **2013**, *61*, 2895–2903. [CrossRef]
4. Matsunaga, N.; Sanada, A.; Kubo, H. Novel Two dimensional Planar Negative Refractive Index Structure. *IEICE Trans. Electron.* **2006**, *9*, 1276–1282. [CrossRef]
5. Best, S.R. A Discussion on the Significance of Geometry in Determining the Resonant Behavior of Fractal and Other Non-Euclidean Wire Antennas. *IEEE Antennas Propag. Mag.* **2003**, *45*, 9–27. [CrossRef]
6. Werner, D.H.; Ganguly, S. An Overview of Fractal Antenna Engineering Research. *IEEE Antennas Propag. Mag.* **2003**, *45*, 38–53. [CrossRef]
7. Dhar, S.; Patra, K. A dielectric resonant-loaded Minkowski fractal-shaped slot loop heptaband antenna. *IEEE Trans. Antennas Propag.* **2015**, *63*, 1521–1525. [CrossRef]
8. Mosallaei, H.; Sarabandi, K. Antenna Miniaturization and Bandwidth Enhancement Using a Reactive Impedance Substrate. *IEEE Trans. Antennas Propag.* **2004**, *52*, 2403–2414. [CrossRef]
9. Yao, J.; Tchafa, F.M.; Jain, A.; Tjuatja, S.; Huang, H. Far-Field Interrogation of Microstrip Patch Antenna for Temperature Sensing without Electronics. *IEEE Sens. J.* **2016**, *16*, 7053–7060. [CrossRef]
10. Le-Wei, L.; Ya-Nan, L.; Soon, Y.T.; Mosig, R.J.; Martin, J.F. A broadband and High Gain Metamaterial Antenna. *Appl. Phys. Lett.* **2010**, *96*, 1–3.
11. Kubacki, R.; Lamari, S.; Czyżewski, M.; Laskowski, D. A broadband left-handed metamaterial microstrip antenna with double-fractal layers. *Int. J. Antennas Propag.* **2017**, *2017*, 1–6. [CrossRef]
12. El-Hameed, A.S.; Salem, D.A.; Hashish, E.A. Crossbar fractal quasi self-complementary UWB antenna. In Proceedings of the 2014 IEEE Antennas and Propagation Society International Symposium (APSURSI), Memphis, TN, USA, 6–11 July 2014; pp. 219–220.
13. Werner, D.H.; Lee, D. A Design Approach for Dual-Polarized Multiband Frequency Selective Surfaces Using Fractal Elements. *IEEE Int. Symp. Antennas Propag.* **2000**, *3*, 1692–1695.
14. Werner, D.H.; Lee, D. Design of Dual-Polarized Multiband Frequency Selective Surfaces Using Fractal Elements. *IEEE Electr. Lett.* **2000**, *36*, 487–488. [CrossRef]
15. Moraes, L.B.; Barbin, S.E. A comparison between Minkowski and Koch Fractal Patch antennas. In Proceedings of the 2011 SBMO/IEEE MTT-S International Microwave & Optoelectronics Conference (IMOC), Natal, Brazil, 29 October–1 November 2011; pp. 17–21.
16. Shafe, S.N.; Adam, I.; Soh, P.J. Design and simulation of a modified Minkowski fractal antenna for tri-band application. In Proceedings of the Fourth Asia International Conference on Mathematical/Analytical Modelling and Computer Simulation, Bornea, Malaysia, 26–28 May 2010; pp. 567–570.

17. Singh, A.; Kumar, M. Design and Simulation of Miniaturized Minkowski Fractal Antennas for microwave applications. *Int. J. Adv. Res. Comput. Commun. Eng.* **2014**, *3*, 5309–5311.
18. Comisso, M. On The Use Of Dimension And Lacunarity for Comparing the Resonant Behavior of Convoluted Wire Antennas. *Prog. Electromagnet. Res.* **2009**, *96*, 361–376. [CrossRef]
19. Gianvittorio, J.P.; Rahmat-Samii, Y. Fractal Antennas: A Novel Antenna Miniaturization technique and Application. *IEEE Antennas Propag. Mag.* **2002**, *44*, 20–36. [CrossRef]
20. Comisso, M. Theoretical and Numerical Analysis of the Resonant Behaviour of the Minkowski Fractal Dipole Antenna. *IET Microw. Antennas Propag.* **2009**, *3*, 456–464. [CrossRef]
21. Best, S.R. A Comparison of the Performance Properties of the Hilbert Curve Fractal and Meander Line Monopole Antennas. *Microw. Opt. Technol. Lett.* **2002**, *35*, 258–262. [CrossRef]

![applied sciences logo] *applied sciences*

MDPI

Article

Dispersion Properties of an Elliptical Patch with Cross-Shaped Aperture for Synchronized Propagation of Transverse Magnetic and Electric Surface Waves

Amagoia Tellechea [1,*], Iñigo Ederra [1,2], Ramón Gonzalo [1,2] and Juan Carlos Iriarte [1,2]

[1] Campus Arrosadía, Public University of Navarra, 31006 Pamplona, Navarra, Spain;
    inigo.ederra@unavarra.es (I.E.); ramon@unavarra.es (R.G.); jcarlos.iriarte@unavarra.es (J.C.I.)
[2] Campus Arrosadía, Institute of Smart Cities, 31006 Pamplona, Navarra, Spain
*   Correspondence: amagoia.tellechea@unavarra.es; Tel.: +34-948-166044

Received: 28 February 2018; Accepted: 16 March 2018; Published: 19 March 2018

Abstract: This paper presents a novel pixel geometry for the implementation of metasurfaces requiring synchronized phase propagation of transverse magnetic (TM) and transverse electric (TE) modes. The pixel is composed by an elliptical metallic patch with an asymmetric cross-shaped aperture in the center, printed on a grounded slab. A practical implementation of a metasurface was carried out employing such a pixel geometry. Simulation results show similar frequency dispersion properties for both modes within the working frequency band, in agreement with the theoretical basis.

Keywords: metasurfaces (MTSs); transverse magnetic (TM); transverse electric (TE); surface waves (SWs); dispersion

---

1. Introduction

Metasurfaces (MTSs) are composed of a dense layer of subwavelength metallic elements, also called pixels [1–3], printed on top of a grounded substrate. Bounded slow surface waves (SWs) can be guided on these structures with engineered dispersion properties. This is the case for the near-field plates, planar lenses and cloaking structures found in the literature [4–7]. Based on the transverse resonance condition, such structures have been characterized by means of scalar or tensor surface impedances [1]. In general, scalar impedance surfaces can support the propagation of either TM or TE SWs. For simplicity, efforts have been focused on guiding of the single TM mode. In fact, several studies have been carried out to analytically characterize the dispersion properties of the TM mode within circular, elliptical and slotted patches printed on top of a grounded slab [8–12].

However, one of the most interesting application of MTS-based structures is the realization of low-profile lightweight antennas for space applications [13–18]. By modulating the characteristic surface impedance, a transition from a cylindrical SW excited form the center to a leaky-wave (LW) is obtained using a MTS-based structure. By means of the modulation of the scalar or tensor impedance characteristics of the isotropic or anisotropic surfaces, beam pointing, shape and radiated field polarization can be tailored for different applications [13–18]. Most of the configurations found in the literature work in single-mode operation. In fact, a single TM SW is supported in such structures, and consequently, single circular polarization (CP) performance is obtained. In order to obtain dual circularly polarized broadside radiation, two modes of TM and TE nature need to be supported by the MTS [19,20]. In these solutions, both modes must have decoupled propagation, be balanced in amplitude and phase synchronized. When these modes interact with a properly modulated surface impedance tensor, a LW radiates outside the structure towards the broadside with the right-hand (RH) or left-hand (LH) CP, depending on the feed phase. The main goal of this work is to discuss a practical pixel geometry supporting the propagation of TM and TE modes balanced in phase that can be employed to implement dual circularly polarized MTS antennas.

The paper is structured as follows. The theoretical principle for phase-matched dual-mode operation is summarized in Section 2. A pixel geometry that meets this condition is presented in Section 3, where a full-wavefrequency and spatial-dispersion analysis is carried out. Additionally, a phase-balanced single-layer guiding MTS is implemented with these pixels. In Section 4, the simulation results are shown. Finally, conclusions are drawn in Section 5.

2. Theoretical Principle

As it is well known [4–17], guiding metasurfaces (MTSs) support the propagation of surface waves (SWs) characterized by a phase velocity ($v = \omega/\beta$) lower than the free space velocity ($c = \omega/k_0$). Thus, the phase constant of the SWs (β) is larger than the free space wavenumber (k_0): $v < c \rightarrow \beta > k_0$. When a cylindrically symmetric MTS is in the xy plane and these modes are excited from the center of the structure, they propagate in a radial direction, bounded to the surface, following $e^{-j\beta\rho}$ (with $\hat{\rho}$ being the radial unit vector). Away from the structure, the field is evanescent, and it is defined by an exponential decay $e^{-jk_{zo}z}$, fulfilling: $k_0^2 = \beta^2 + k_{zo}^2$, where $\Im(k_{zo}) = -\alpha_{zo}$, $\alpha_{zo} > 0$.

The goal of this work is to implement a single-layer metasurface that supports the propagation of a transverse magnetic mode (TM) and a transverse electric mode (TE). As has been widely explained in [8–13], a single-layer MTS can be analyzed in terms of its equivalent circuit model. According to the transverse resonance equation, the surface impedance relating the fields of each SW in the surface ($Z_{surf,i}$) can be related to the free space impedance of each mode at the air-surface interface ($Z_{0,i}$) as: $Z_{0,i\ i=TM,TE} = -Z_{surf,i\ i=TM,TE}$. Thus, the surface impedance of each mode can be described by the following expressions:

$$Z_{surf,TM} = j\,X_{surf,TM} = -\zeta\frac{k_{zo}}{\beta} \tag{1}$$

$$Z_{surf,TE} = j\,X_{surf,TE} = -\zeta\frac{\beta}{k_{zo}} \tag{2}$$

where $\zeta = 120\pi$ is the free space impedance. From Equations (1) and (2), it can be concluded that the surface impedance of the desired MTS must have an inductive reactive term in order to allow the propagation of a TM SW ($X_{surf,TM} > 0$), while the surface impedance must have an inductive reactive term ($X_{surf,TE} < 0$), to support the TE mode propagation.

Both modes must propagate in the structure decoupled and synchronized in phase. The condition for TM and TE modes phase synchronism can also be related to the equivalent surface reactance values as:

$$\beta_{TM} = \beta_{TE} \rightarrow X_{surf,TM}\,X_{surf,TE} = -\zeta^2 \tag{3}$$

Exciting the rotationally symmetric MTS from the center by a magnetic dipole (slot in the ground plane in a practical implementation) the TM and TE modes propagate radially from the center. In the MTS configuration shown in Figure 1, the magnetic dipole (depicted as a black arrow) is oriented towards the y-axis. In this case, the TM SW is azimuthally excited following a $\cos(\varphi)$ function, while the field related to the TE mode meets a $\sin(\varphi)$ excitation. If the phase synchronism of the modes is ensured all over the surface and the mode amplitudes are equal, the total field on the surface, obtained as the combination of both modes, provides a linearly polarized E_x field:

$$E^{(y)} = E_{TM}^{\ (y)} + E_{TE}^{\ (y)} = -E_{TM}^{\ (y)}\cos(\varphi)\hat{\rho} + E_{TE}^{\ (y)}\sin(\varphi)\hat{\varphi} = -E_0\hat{x} \tag{4}$$

where $E_{TM} = E_{TE} = E_0$ for circular polarization.

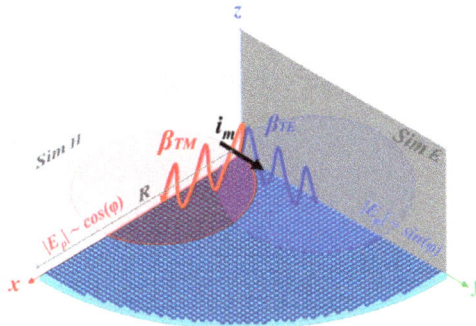

Figure 1. A rotationally symmetric metasurface is excited by a magnetic dipole oriented towards the *y*-axis: TM and TE SWs propagate radially on the surface phase synchronized. $|E\rho|$ follows a $\cos(\varphi)$ function while $|E\varphi|$ follows a $\sin(\varphi)$ function.

3. Practical Implementation

3.1. Unit Cell Geometry

The geometry of the proposed pixel is shown in Figure 2. An elliptically shaped metallic patch with a cross-shaped aperture is printed on top of a grounded dielectric of AD-1000 material, with a relative permittivity of $\varepsilon_r = 10.2$ and a thickness of $h = 1.27$ mm. The unit cell dimension is $u = 3.14$ mm. The minor axis of the elliptical metallic patch (e_a) is oriented towards the *x*-axis and the major axis (e_b) towards the *y*-axis in the rectangular reference system. Inside the metallic patch, there is an asymmetric cross-shaped aperture. The main axes of this cross aperture are also aligned with the rectangular reference, i.e., they are aligned with the ellipse axes. The length and width of the cross are denoted as c_a, w_a and c_b, w_b towards the *x* and *y* axis, respectively. Since the main applications of the MTSs composed with such subwavelength elements are in the microwave regime, the metal is treated as a perfect electric conductor, and dielectric losses are neglected.

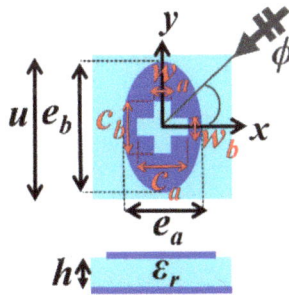

Figure 2. Unit cell geometry composed by a metallic elliptical patch with asymmetric cross-shaped aperture in the center, printed on a grounded substrate. Dimension $u = 3.14$ mm, AD-1000 dielectric with relative permittivity $\varepsilon_r = 10.2$ and thickness $h = 1.27$ mm.

A full wave eigenmode analysis for the supported SWs has been carried out. For this purpose, periodic boundary conditions are imposed in the *x* and *y* axis with $k_x u$ and $k_y u$ phasing, respectively. The imposed phasing corresponds to a mode impinging on the cell towards the ϕ angle, which is evanescent in the *z* direction. Thus, it meets: $k_x^2 + k_y^2 = \beta_i^2 > k_0^2$.

3.2. Pixel Dispersion Properties

The transverse magnetic (TM) and transverse electric (TE) modes' spatial and frequency dispersion properties for different patches are shown and compared in this section.

In order to facilitate reader understanding, firstly the TM and TE modes' isofrequency dispersion curves for a pixel composed by an elliptical metallic patch printed on a grounded dielectric (i.e., without cross slot) are shown in Figure 3. The dispersion properties for different elliptical patch ratios (e_a/e_b) are compared.

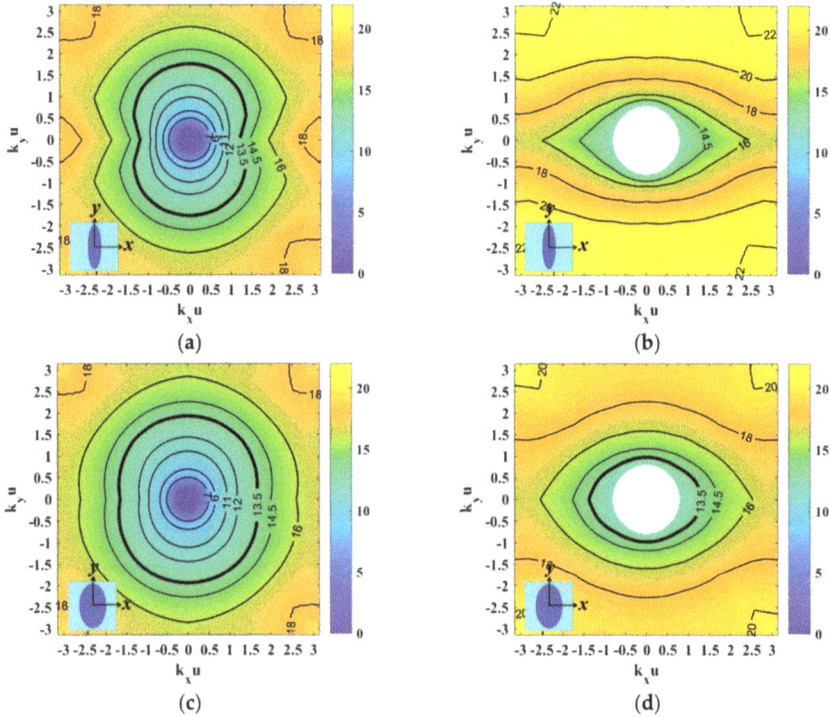

Figure 3. Frequency and spatial dispersion curves for TM (left column) and TE (right column) SWs supported by a pixel composed by a metallic elliptical patch printed on a grounded dielectric with different ratios: (**a,b**) e_a/e_b = 0.6 and (**c,d**) e_a/e_b = 0.85. The employed dielectric is AD-1000 (ε_r = 10.2, thickness h = 1.27 mm) and the unit cell dimension u = 3.14 mm.

As can be seen in Figure 3a, the phase constant of the supported TM mode (β_{TM}) increases when the electrical size of the patch is larger. The TM phase constant is always larger when the propagation angle is oriented towards the larger axis of the ellipse (in this case the y-axis), and the same effect can be observed in the x-axis when the ratio of the ellipse is increased (Figure 3c). Unlike the TM mode, which is supported in the structure at all frequencies, the TE SW has a cut-off frequency. As can be seen in Figure 3b–d, the cut-off frequency of the TE mode decreases for electrically large patches. For instance, in the patch with larger ellipse ratio (Figure 3d), the TE mode is outside the visible region at 13.5 GHz (shown in bold in the isofrequency curves), while it is inside the visible region for smaller ratios (Figure 3b). The TE phase constant (β_{TE}) increases on the x-axis for ellipses with larger ratios.

When the required MTS must support only the propagation of a TM mode, the geometry and rotations of the selected patches for the MTS implementation must be carefully selected to avoid the propagation of the TE SW working below its cut-off frequency. Nevertheless, in the present work, both the TM and TE modes must be considered. The main goal is to independently control the propagation properties of both the TM and TE SWs, ensuring, additionally, the synchronized propagation of both modes within a certain impinging angle (ϕ). For this purpose, an asymmetric cross-shaped aperture has been introduced in the center of the previously presented elliptical patch geometry (shown in Figure 2). Frequency and spatial dispersion properties for three geometric parameterizations are shown in Figure 4.

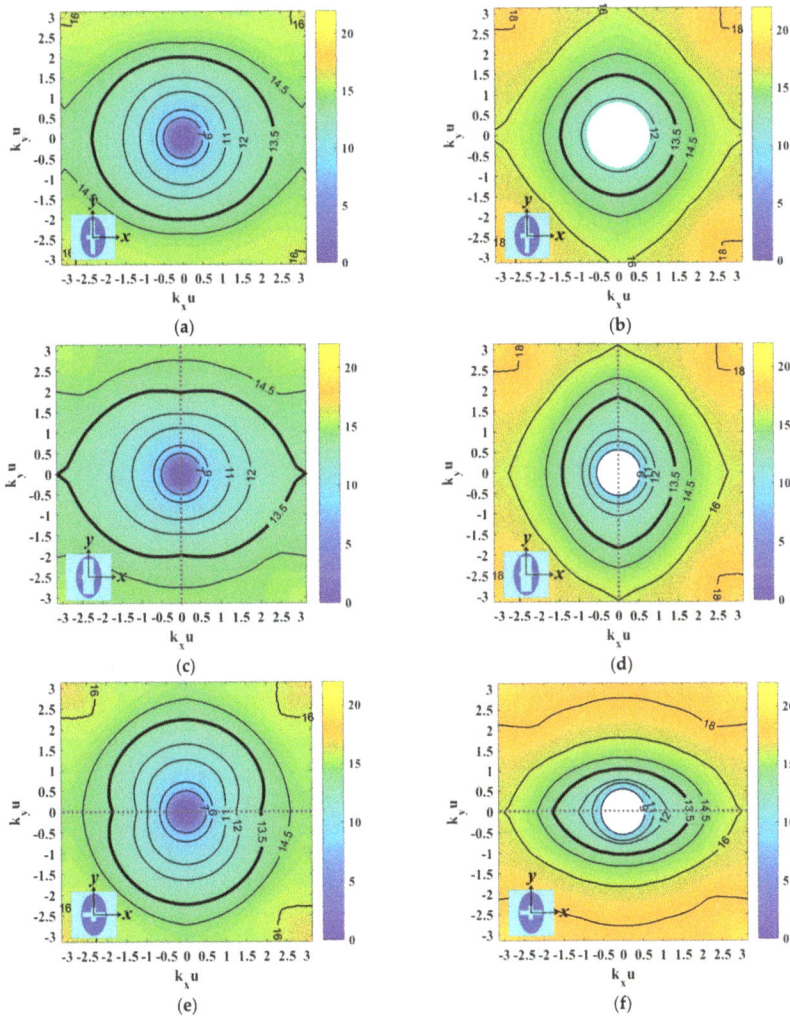

Figure 4. Frequency and spatial dispersion curves for TM (left column) and TE (right column) SWs supported by three different pixels composed by a metallic elliptical patch with an asymmetric aperture in the center, printed on a grounded dielectric. The ellipse ratio for the patches is $e_a/e_b = 0.85$ in all cases, and cross apertures are defined by (w_a/u, c_a/e_a, w_b/u, c_b/e_b) parameters: (**a,b**) pixel A (0.15, 0.6, 0.15, 0.9), (**c,d**) pixel B (0.25, 0.6, 0.15, 0.9) and (**e,f**) pixel C (0.15, 0.85, 0.15, 0.6). The employed dielectric is AD-1000 ($\varepsilon_r = 10.2$, thickness $h = 1.27$ mm) and the unit cell dimension $u = 3.14$ mm.

Given that the unit cell dimension is very small in terms of wavelength, the inclusion of the asymmetric cross-shaped aperture in the center of the elliptical patch geometry makes it possible to control the phase constant of the different modes (β_{TM}, β_{TE}) independently when impinging on the unit cell with different angles. TM mode dispersion properties are affected by the slot of the cross perpendicular to the impinging angle, while the TE SW is affected by the slot aligned with the incidence direction. In Figure 4a, the cross slot along the larger axis of the elliptical patch (y-axis) has been enlarged, increasing the c_b parameter. Due to this fact, when a TM mode propagates in the unit cell along the x-axis, β_{TM} increases considerably. Increasing c_b also affects the dispersion properties of the TE mode (Figure 4b); when the mode propagates in the cell towards $\phi = 90°$, β_{TE} increases in the y-axis. On the other hand, the effect of broadening the slot corresponds to an electrically larger patch. In Figure 4c,d, the slot width towards the x-axis (w_a parameter) is enlarged, and due to this effect, β_{TM} and β_{TE} increase when TM and TE modes propagate in the cell at angle of $\phi = 0°$ and $\phi = 90°$, respectively.

It is necessary to remark that this patch geometry is especially attractive due to the multiple possibilities for its geometrical parameterization, which make it possible to independently control the dispersion properties for both TM and TE modes within a certain frequency band. For instance, the patch geometry shown in Figure 4c,d (pixel B) allows the synchronized propagation of TM and TE SWs when impinging on the patch along its larger axis (in this case, $\phi = 90°$) at 13.5 GHz. As can be seen, the isofrequency curves of both modes coincide at this frequency, and the phase constant of both modes is the same: $\beta_{TM} = \beta_{TE} = 597$ rad/m. However, the modes' synchronization can be also obtained with other geometrical parameterization. This is the case, for instance, for pixel C (shown in Figure 4e,f), in which the cross slot is increased along the shorter axis of the ellipse (c_a parameter has been increased). This geometry also ensures $\beta_{TM} = \beta_{TE} = 597$ rad/m at 13.5 GHz, when TM and TE modes impinge on the cell along the $\phi = 0°$ angle. This feature is also shown in Figure 5.

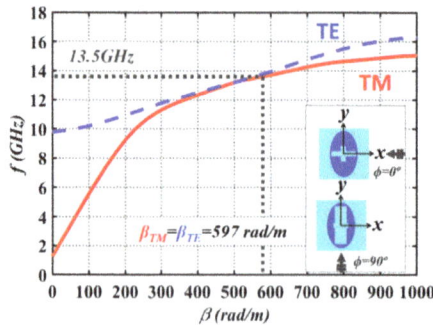

Figure 5. Frequency dispersion properties for TM and TE modes when impinging on pixel B at $\phi = 0°$. Curves agree with the frequency dispersion of the modes for pixel C within $\phi = 90°$ angle.

It has to be mentioned that the phase synchronization bandwidth is mainly limited by the TE mode (see Figure 5), as the frequency dispersion behavior of this mode changes faster with frequency than the TM SW dispersion.

3.3. Metasurface Implementation Details

In this section, the implementation details of a SW-guiding MTS are given, in which two TM and TE modes are guided, decoupled and phase-balanced (Figure 6). Pixel B, presented in the previous Section 3.2, has been employed to implement a SW-guiding surface with a total radius of $R = 50u = 157$ mm $= 7\lambda_0$. At each position on the surface, the metallic elliptical patch is printed on top of the grounded dielectric, based on a cartesian lattice meshgrid (unit cell dimension $u = 3.14$ mm). Each patch is appropriately rotated, in such a way that the minor axis of the ellipse meets the impinging angle of the SWs propagating from the center of the structure, i.e., is aligned with $\hat{\rho}$. The structure is excited by a magnetic dipole oriented towards the y-axis.

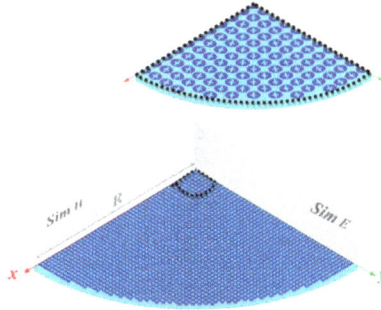

Figure 6. Details of the rotationally symmetric metasurface composed by pixel B subwavelength elements. Total radius $R = 50u = 157$ mm. The structure is excited by a magnetic dipole oriented towards the y-axis. The metal patches are shown in dark blue, whereas the light blue color represents the dielectric substrate.

4. Simulation Results

Employing appropriate symmetry conditions (see Figure 6), a quarter of the MTS structure was simulated using ANSYS HFSS software package. Two TM and TE modes are excited by a magnetic dipole oriented towards the y-axis. The total near-field components at 13.5 GHz obtained by the combination of both modes guided on the structure are depicted in Figure 7.

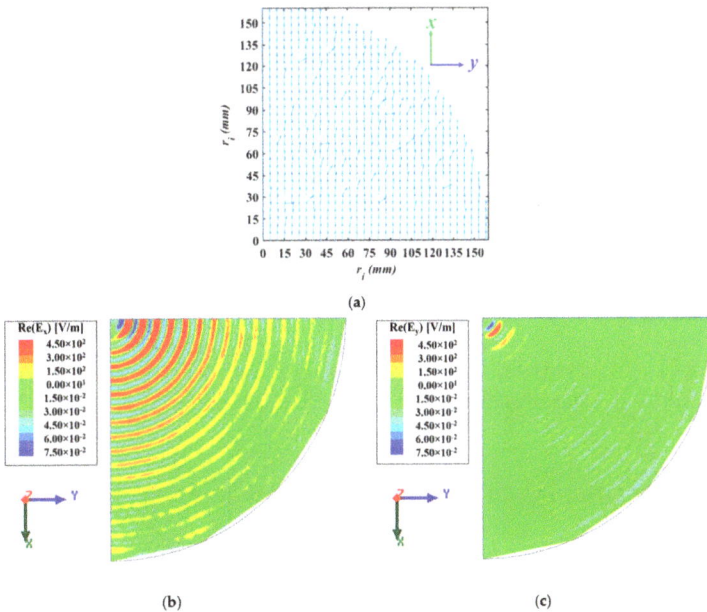

Figure 7. (a) Total field vector at 13.5 GHz over the MTS obtained as the combination of TM and TE modes; (b) $Re(E_x)$ and (c) $Re(E_y)$ components' magnitude over the surface.

The TM and TE modes were appropriately excited on the structure, with propagation that was synchronized and in-phase. Due to this fact, the TM and TE modes' field combination generates a total field oriented towards the *x*-axis (Figure 7a). Additionally, the decoupled propagation of both modes ensures negligible cross polar field excitation. In consequence, $Re(E_y)$ is negligible in Figure 7c.

Figure 8 depicts the total $Re(E_x)$ field in the aperture at $\varphi = 0°$, $45°$ and $90°$ planes, obtained as a combination of the synchronized TM and TE SW fields.

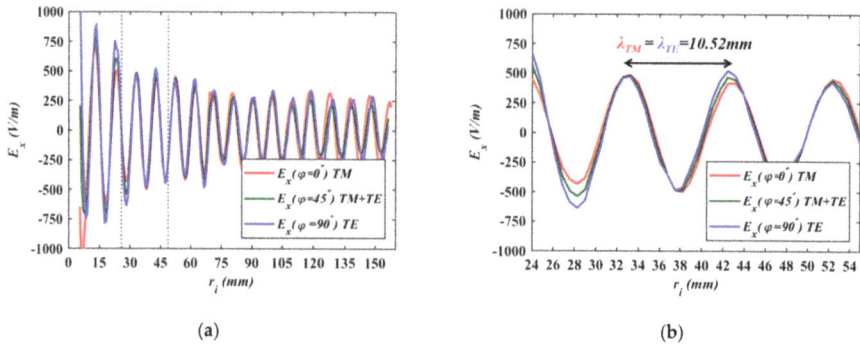

(a)   (b)

Figure 8. (a) $Re(E_x)$ field in the aperture ($R = 50u = 157$ mm) at $\varphi = 0°, 45°, 90°$ as a combination of the TM and TE SW fields; (b) Zoom of the curves.

As expected, the SW wavelengths of both modes are in agreement, $\lambda_{TM} = \lambda_{TE} = 10.52$ mm (Figure 8b). This feature is in agreement with the predicted phase velocity values related to the unit cell dispersion analysis carried out in Section 3.2, where $\beta_{TM} = \beta_{TE} = 597$ rad/m. It has to be noted that the synchronism of both modes is perfectly maintained in the center of the antenna. Nevertheless, for the outer positions ($r_i > 5.5\lambda_0$), there is some phase difference between the modes. The reason for this effect may be related to the cartesian lattice for the surface implementation, and its effect on the TE dispersion properties when the patch is rotated azimuthally.

5. Conclusions

Design details of a single-layer rotationally symmetric metasurface ensuring decoupled and synchronized propagation of TM and TE modes has been presented in this work. The MTS was implemented by means of a dense layer of subwavelength metallic elliptical patches with an asymmetric cross-shaped aperture in the center, printed on top of a grounded dielectric. With such a pixel geometry, the dispersion properties of TM and TE SWs match at the working frequency (13.5 GHz), and the total field generated on the aperture by the combination of both modes is perfectly linearly polarized. The design of such structures has important applications in the development of dual circularized polarized MTS antennas based on surface impedance modulation, in which TM and TE mode propagation with similar dispersion properties is required.

Acknowledgments: The authors would like to thank the financial support by the Spanish Ministry of Economy and Competitiveness, Project No. TEC2013-47753-C3-1-R and TEC2016-76997-C3-1-R and by the Gobierno de Navarra, Project PI017 COMUNICACIONES 5G and PI018 COMUNICACIONES 5G.

Author Contributions: Amagoia Tellechae and Juan Carlos Iriarte conceived and designed the experiments; Amagoia Tellechea performed the experiments; Iñigo Ederra y Ramón Gonzalo analyzed the data; Amagoia Tellechea, Iñigo Ederra y Juan Carlos Iriarte wrote the paper.

Conflicts of Interest: The authors declare no conflict of interest.

Appl. Sci. **2018**, *8*, 472

References

1. Fong, B.H.; Colburn, J.S.; Ottusch, J.J.; Visher, J.L.; Sievenpiper, D.F. Scalar and tensor holographic artificial impedance surfaces. *IEEE Trans. Antennas Propag.* **2010**, *58*, 3212–3221. [CrossRef]
2. Bilow, H.J. Guided waves on a planar tensor impedance surface. *IEEE Trans. Antennas Propag.* **2003**, *51*, 2788–2792. [CrossRef]
3. Sievenpiper, D.; Colburn, J.; Fong, B.; Ottusch, J.; Visher, J. Holographic artificial impedance surfaces for conformal antennas. In Proceedings of the 2005 IEEE Antennas and Propagation Society International Symposium, Washington, DC, USA, 3–8 July 2005.
4. Jiang, L.; Grbic, A.; Merlin, A. Near-field plates: Subdiffraction focusing with patterned surfaces. *Science* **2008**, *320*, 511–513.
5. Pfeiffer, P.; Grbic, A. Planar lens antennas of subwavelength thickness: Collimating leaky-waves with metasurfaces. *IEEE Trans. Antennas Propag.* **2015**, *63*, 3248–3253. [CrossRef]
6. Martini, E.; Mencagli, M.; Maci, S. Metasurface transformation for surface wave control. *Philos. Trans. R. Soc. A* **2015**, *373*. [CrossRef] [PubMed]
7. González-Ovejero, D.; Martini, E.; Maci, S. Surface waves supported by metasurfaces with self-complementary geometries. *IEEE Trans. Antennas Propag.* **2015**, *63*, 250–260. [CrossRef]
8. Patel, A.M.; Grbic, A. Modeling and analysis of printed-circuit tensor impedance surfaces. *IEEE Trans. Antennas Propag.* **2013**, *61*, 211–220. [CrossRef]
9. Mencagli, M.; Martini, E.; Maci, S. Surface wave dispersion for anisotropic metasurfaces constituted by elliptical patches. *IEEE Trans. Antennas Propag.* **2015**, *63*, 2992–3003. [CrossRef]
10. Granet, G.; Luukkonen, O.; Simovski, C.; Tretyakov, S.A. Simple and accurate analytical model of planar grids and high-impedance surfaces comprising metal strips or patches. *IEEE Trans. Antennas Propag.* **2008**, *56*, 1624–1632.
11. Mencagli, M.; Martini, E.; Maci, S. Transition function for closed-form representation of metasurface reactance. *IEEE Trans. Antennas Propag.* **2016**, *64*, 136–145. [CrossRef]
12. Mencagli, M.; Giovampaola, C.D.; Maci, S. A closed-form representation of isofrequency dispersion curve and group velocity for surface waves supported by anisotropic and spatially dispersive metasurfaces. *IEEE Trans. Antennas Propag.* **2016**, *64*, 2319–2327. [CrossRef]
13. Oliner, A.; Hessel, A. Guided waves on sinusoidally-modulated reactance surfaces. *IRE Trans. Antennas Propag.* **1959**, *7*, 201–208. [CrossRef]
14. Minatti, G.; Maci, S.; De Vita, P.; Freni, A.; Sabbadini, M. A circularly-polarized isoflux antenna based on anisotropic metasurface. *IEEE Trans. Antennas Propag.* **2012**, *60*, 4998–5009. [CrossRef]
15. Minatti, G.; Faenzi, M.; Martini, E.; Caminita, F.; De Vita, P.; González-Ovejero, D.; Sabbadini, M.; Maci, S. Modulated metasurface antennas for space: Synthesis, analysis and realizations. *IEEE Trans. Antennas Propag.* **2015**, *63*, 1288–1300. [CrossRef]
16. Minatti, G.; Caminita, F.; Casaletti, M.; Maci, S. Spiral leaky-wave antennas based on modulated surface impedance. *IEEE Trans. Antennas Propag.* **2011**, *59*, 4436–4444. [CrossRef]
17. Faenzi, M.; Caminita, F.; Martini, E.; de Vita, P.; Minatti, G.; Sabbadini, M.; Maci, S. Realization and measurement of broadside beam modulated metasurface antennas. *IEEE Antennas Wirel. Propag. Lett.* **2016**, *15*, 610–613. [CrossRef]
18. Gonzalez-Ovejero, D.; Minatti, G.; Chattopadhyay, G.; Maci, S. Multibeam by metasurface antennas. *IEEE Trans. Antennas Propag.* **2017**, *65*, 2923–2930. [CrossRef]
19. Tellechea, A.; Caminita, F.; Martini, E.; Ederra, I.; Iriarte, J.C.; Gonzalo, R.; Maci, S. Dual circularly polarized broadside beam metasurface antenna. *IEEE Trans. Antennas Propag.* **2016**, *64*, 2944–2953. [CrossRef]
20. Tellechea, A.; Iriarte, J.C.; Ederra, I.; Gonzalo, R.; Martini, E.; Maci, S. Experimental validation of a Ku-band dual circularly polarized metasurface antenna. *IEEE Trans. Antennas Propag.* **2018**, *66*, 1153–1159.

![applied sciences logo] *applied sciences*

MDPI

Article

Design of 1-Bit Digital Reconfigurable Reflective Metasurface for Beam-Scanning

Shuncheng Tian, Haixia Liu * and Long Li *

Key Laboratory of High Speed Circuit Design and EMC of Ministry of Education, School of Electronic Engineering, Collaborative Innovation Center of Information Sensing and Understanding, Xidian University, Xi'an 710071, China; sctian@xidian.edu.cn
* Correspondence: hxliu@xidian.edu.cn (H.L.); lilong@mail.xidian.edu.cn (L.L.); Tel.: +86-029-8820-1157 (L.L.)

Received: 1 August 2017; Accepted: 25 August 2017; Published: 28 August 2017

Abstract: A 1-bit digital reconfigurable reflective metasurface (RRM) with 20×20 cells is presented, fabricated and measured for beam-scanning performance in this paper. The cell is designed with a single layer structure and one varactor diode, controlled electronically. The cell's phase compensation is over 180° between 3 GHz and 4 GHz and the two states with 180° phase difference are selected as coding "0" and coding "1". By the fuzzy quantification theory, all the elements on the RRM are set to be coding "0" or coding "1" according to the phase compensation calculated by MATLAB. Furthermore, by changing the coding of the RRM, it can achieve beam-scanning. The simulation results show that the beam-scanning range is over ±60°. The RRM prototype is fabricated and experimentally tested for principle. The gain of the RRM is 18 dB and the 3 dB bandwidth is about 16.6%. The 1-bit digital RRM is preferred in practical implementations due to less error and much easier bias voltage control. The proposed RRM successfully balances the performance and system complexity, especially in the large-scale antenna designs. The experimental and simulated results are in good agreement to prove the correctness and feasibility of the design of the 1-bit digital RRM.

Keywords: 1-bit; digital; RRM; varactor diode; beam-scanning

1. Introduction

Due to its hybrid phased array and parabolic reflector array, reconfigurable reflective metasurface (RRM) has attracted a great deal of attention and interest recently [1]. With electronic switches, PIN diodes, metamaterials and so on, RRM can produce versatile performances, for instance, lightweight, low profile, less system complexity, high efficiency and flexible radiation characteristics [2,3]. Consequently, RRM have been proved to be an up-and-coming alternative to traditional antenna in satellite communications, long distance transmissions, imaging and in other realms [4].

In some research, PIN diode is adopted to design a metasurface element [3,5,6]. Since the PIN diode only works at one frequency at two states, ON and OFF, the bandwidth of the reconfigurable reflective metasurface is very narrow. If we want to change the operating frequency to get a wide bandwidth, the range of the cell's phase compensation maybe not adequate. When the frequency changes, the element may not be appropriate to design the reflective metasurface [7,8]. However, the element with varactor diode can work at different frequencies by controlling the bias voltage across the varactor. As a result, it has more potential applications since it can work in a wider bandwidth [9–11]. These electronically-controlled digital reflective metasurfaces have been proposed for different practical applications such as tunable absorbers, reconfigurable antennas, EM wave modulators, tunable chiral metamaterials, reprogrammable hologram, and so on [12,13].

In this paper, a 1-bit digital RRM is designed, fabricated, and experimentally demonstrated to achieve beam-scanning. The 1-bit digital reconfigurable reflective metasurface unit's phase compensation is over 180° between 3 GHz and 4 GHz and the two states with 180° phase difference are selected as coding "0" and coding "1". In addition, a 1-bit digital RRM composed of 20 × 20 units is presented. Due to the theory of fuzzy quantification, every element on the RRM is set to be coding "0" or coding "1" to replace the desired phase compensation [14,15]. The RRM realizes the beam-scanning performance by controlling the coding and the scanning angle can reach ±60°. Finally, the antenna has been manufactured and measured to ensure that the design is credible and feasible. The results of this research provide new perspectives on the use of the digital reprogrammable reconfigurable reflective metasurface.

2. Materials and Methods

2.1. Description of the 1-Bit Digital Element and RRM

In this study, the 1-bit digital reconfigurable reflective metasurface element is designed to obtain the digital phase compensation to replace the accurate phase compensation. As shown in Figure 1a, the 1-bit digital reconfigurable reflective metasurface element adopts a single layer patch structure with a varactor diode through the gap between the two patches. The unit is etched on a dielectric substrate and the bottom is the metal ground. The dielectric constant of the dielectric substrate is 2.65 and the thickness is 3 mm.

The remaining structure parameters of the 1-bit digital element are provided in Table 1. The proposed element is simulated with the infinite periodic boundary condition in HFSS 15.0 [16]. By controlling the bias voltage across varactor diode, the phase compensation of the element can be adjusted as desired. A 1-bit 20 × 20 RRM is present and fabricated at 3.5 GHz fed by a horn antenna using the above element, as shown in Figure 1b,c. The size of the RRM is 500 mm × 500 mm. Figure 1d gives the diagram of incident wave, reflected wave, incident angle and reflection angle in the study. The incident wave is in the *xoz* plane, and the reflection wave is in the *yoz* plane.

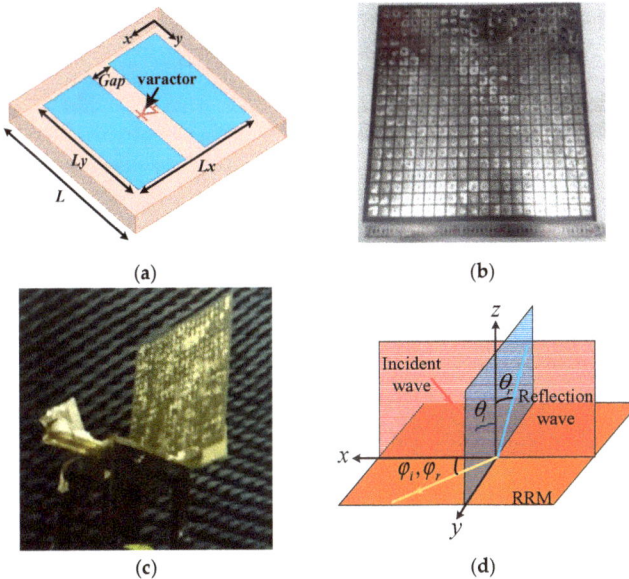

Figure 1. Geometry of (**a**) The 1-bit digital element; (**b**) The fabricated reconfigurable reflective metasurface (RRM); (**c**) The experimental scene; (**d**) The incident wave, reflected wave, incident angle and reflection angle.

Table 1. Parameters of the 1-bit digital RRM element.

Parameter	Value	Parameter	Value
L	25 mm	Lx	22.5 mm
Gap	0.5 mm	Ly	22.5 mm

2.2. The Theory of Reflective Metasurface

An RRM, which consists of $m \times n$ elements, is illuminated by a feed source locating at the position vector $\vec{r}_f$. The reradiated electric field from the RRM in an arbitrary direction can be calculated by [2]

$$E(\hat{u}) = \sum_{m=1}^{M} \sum_{n=1}^{N} F(\vec{r}_{mn} \cdot \vec{r}_f) A(\vec{r}_{mn} \cdot \hat{u}_0) A(\hat{u}_0 \cdot \hat{u}) \cdot \exp(-jk_0(|\vec{r}_{mn} - \vec{r}_f| - \vec{r}_{mn} \cdot \hat{u}) + j\alpha_{mn}) \quad (1)$$

where $\hat{u}_0$ is the vector of the main beam direction. F is the function of the radiation pattern of the feed, and A is the function of the radiation pattern of the array element. $\vec{r}_{mn}$ is the position vector of each element, and α_{mn} is the phase compensation of each cell. The phase compensation of each RRM element for a desired direction can be computed by

$$-jk_0(|\vec{r}_{mn} - \vec{r}_f| - \vec{r}_{mn} \cdot \hat{u}) + j\alpha_{mn} = 2n\pi(n = 1, 2, 3 \ldots) \quad (2)$$

When the phase compensations of all units can meet (1) and (2), the RRM can radiate in the required direction. The direction of the incidence wave (θ_i, φ_i) is $(30°, 0°)$ and the direction of the reflection wave (θ_r, φ_r) is $(\theta_r, 90°)$, where θ_r can vary from $-60°$ to $+60°$.

2.3. The Theory of Fuzzy Quantification

Depending on the fuzzy quantification theory, the elements on the RRM are divided into two groups, coding "0" and coding "1". In order to get the two states of the 1-bit digital element, we need to quantify the phase compensation. The quantization standard is that if we want to get the K-bit quantization, the number of the phase states is 2^K for 360°. That is to say, K is 1 for a 1-bit digital element and the phase compensation of the unit only needs to be over 180°.

When $\theta_r = 0°$, 30° and 60°, the actual and the digital phase compensation of the 1-bit digital RRM computed by MATLAB 2017a are illustrated in Figures 2–4, respectively. The black blocks are coding "0" and the white blocks are coding "1". Based on the above results, the models of the RRM are determined and simulated by using HFSS 15.0.

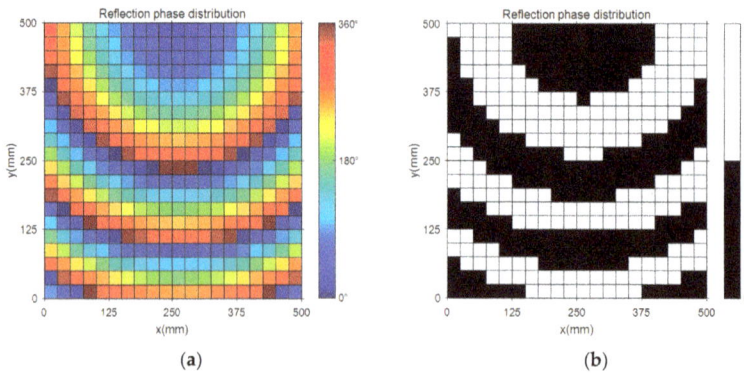

(a)   (b)

Figure 2. Reflection phase distribution on the RRM when $\theta_r = 0°$ (a) Accurate phase distribution; (b) Digital phase distribution.

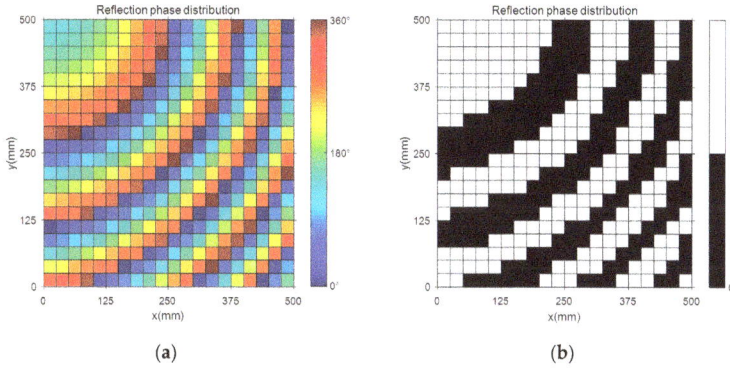

Figure 3. Reflection phase distribution on the RRM when $\theta_r = 30°$ (**a**) Accurate phase distribution; (**b**) Digital phase distribution.

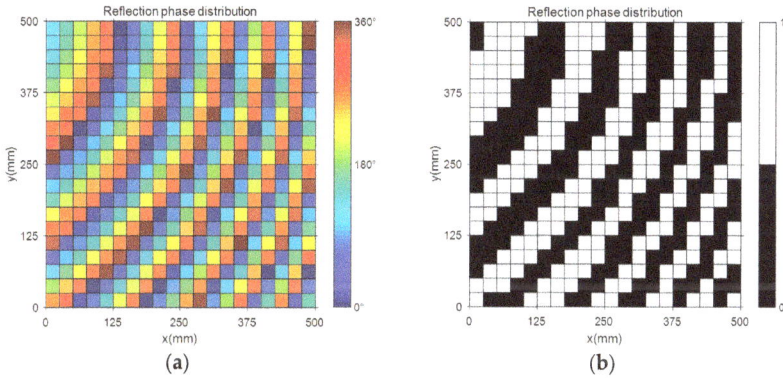

Figure 4. Reflection phase distribution on the RRM when $\theta_r = 60°$ (**a**) Accurate phase distribution; (**b**) Digital phase distribution.

3. Results and Discussion

3.1. Simulation Results of the 1-Bit Digital RRM Element

When the working frequency changes, the 180° phase compensation can be acquired by controlling the capacitance in broadband as given in Figure 5. The varactor diode can alter the phase compensations of all the RRM elements by controlling the bias voltage across the varactor diode. By tuning the capacitance value, the phase compensation curves are plotted in Figure 5a. It can be clearly observed that when the capacitance varies, the range of the phase compensation of the 1-bit digital element is more than 180°. We can see that the element can work between 3 GHz and 4 GHz with over 180° phase compensation. With this characteristic, the RRM can work in a wide bandwidth of 3 GHz to 4 GHz. Therefore, the varactor diode has more potential applications than the PIN diode in the RRM design.

Finally, we select 3.5 GHz as the operating frequency to design the RRM. Based on the simulation result in Figure 5b, the phase compensation is −145° when $Cap = 1.1$ pF and the phase compensation is 35° when $Cap = 1.9$ pF. We select −145° as the coding "0" and 35° as the coding "1". Of course, we can also choose an operating frequency between 3 GHz and 4 GHz to design the 1-bit element by changing the capacitance value.

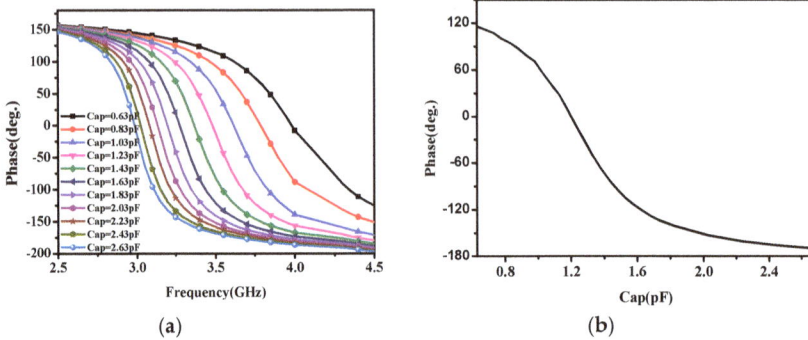

Figure 5. Reflection phase characteristics versus frequencies (**a**) With different capacitance; (**b**) At 3.5 GHz.

Depending on the fuzzy quantification theory, these elements are divided into two groups, corresponding to 35° and −145°. If the phase compensation of the unit is in the range 125~305°, the element is coding "1", and the rest is coding "0". The concept of a digital system means that we transform the phase compensation of a cell into coding "0" and coding "1" of 1-bit. By such a transformation, the RRM generates a digital concept. The digital RRM can realize the beam-scanning performance by changing the coding arrangement. In order to reduce the difficulty of design and processing, we choose the fixed capacitors of $Cap = 1.1\,\text{pF}$ and $Cap = 1.9\,\text{pF}$ to replace the varactor diodes. Because the phase compensations of elements are not calculated by the formulas of reflectarray, this kind of digital phase compensations are not the precise phase compensations. As a result, the gain of the RRM will be less than the current state of the art. In contarst, the digital RRM improves the bandwidth performance.

3.2. Simulation and Experimental Results of the RRM

Figure 6 shows the radiation pattern of the 1-bit digital RRM when the main beam direction is 0°, 30° and 60°, respectively. From the simulation results, the characteristics of beam-scanning have been verified. For the sake of comparing the radiation characteristics in different reflection angles, the far-field performances are given in Table 2. From Table 2, we can see that when the scanning angle increases, the gain of the RRM decreases slightly. It can be seen that the 1-bit digital RRM can radiate in the same reflection angle, as assumed. The far-field radiation patterns are plotted in Figure 7 with different scanning angles and the gain of the RRM decreases slightly with the increase of the reflection angle. Then, by adjusting the coding of the 1-bit digital RRM, it can scan from −60° to +60°. These simulation results verify the validity and practicability of the 1-bit RRM.

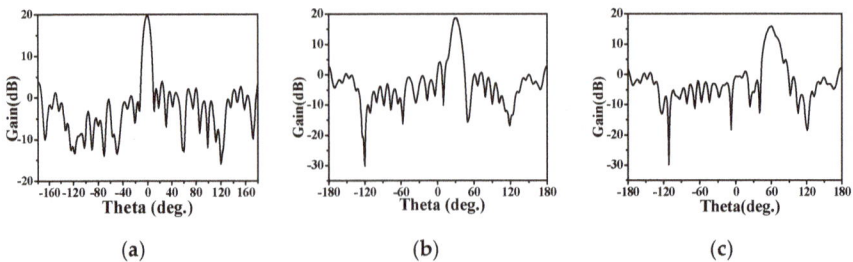

Figure 6. Radiation patterns of the proposed 1-bit digital RRM at (**a**) $\theta_r = 0°$; (**b**) $\theta_r = 30°$; (**c**) $\theta_r = 60°$.

Table 2. The gain of the 1-bit digital RRM at different main beam directions and beam point errors for different scanning angles.

(θ_r, φ_r)	Gain (dB)	Beam Point Error	(θ_r, φ_r)	Gain (dB)	Beam Point Error
$(0°, 90°)$	19.89	0°	$(50°, 90°)$	16.16	0°
$(30°, 90°)$	18.65	0°	$(60°, 90°)$	15.85	0°
$(40°, 90°)$	17.59	0°			

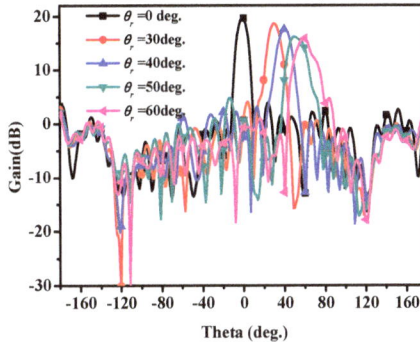

Figure 7. Scanning main beam characteristics of the 1-bit digital RRM.

We test the performances of the proposed 1-bit digital RRM while $(\theta_i, \varphi_i) = (30°, 0°)$ and $(\theta_r, \varphi_r) = (60°, 90°)$ by using fixed capacitors. The simulated and measured results are given in Figure 8. From the simulation results, the 1-bit digital RRM can work from 3.47 GHz to 4.23 GHz, and the relative bandwidth is about 22%. The actual 3 dB bandwidth is 16.6%. The measured gain is slightly decreased. The deviation between the simulation and the measurement results may be caused by the fabrication error and capacitance error. The varactor diode is connected between the two patches by solder, which may lead to the introduction of additional impedance. The welding technique is very important because if the tin solder is too heavy much, the capacitance value of the varactor may not be the same as its original value. The additional impedance is uncontrollable and may result in errors for the measurement results. It is believed that these deficiencies will be readily compensated and thus superior radiation characteristics can be obtained. It is worth pointing out that these experimental results agree well with the simulations results. As a result, if we want to design an RRM with much wider broadband, the digital method maybe a better choice.

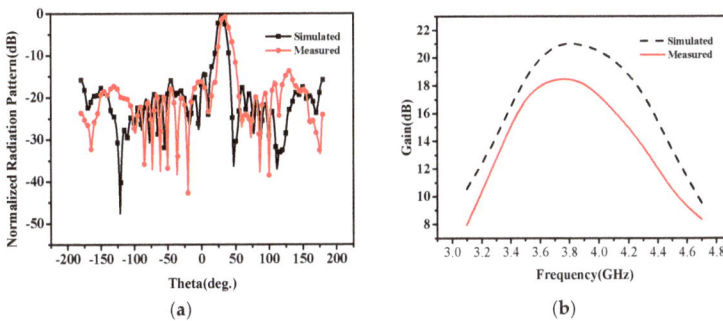

Figure 8. Comparison of measured and simulated results of the fabricated RRM (**a**) Normalized radiation pattern at 3.5 GHz; (**b**) Gain characteristics.

4. Conclusions

In this paper, a novel 1-bit RRM at C-band is simulated and fabricated aiming at achieving the beam-scanning characteristic in wider broadband and decreasing the design difficulty of RRM. The 1-bit element RRM cell, with only one varactor diode, works between 3 GHz and 4 GHz and provides two states with 180° phase difference as coding "0" and coding "1". Then, we design a RRM working at 3.5 GHz. The measured gain of the proposed RRM is 18 dB and 3 dB bandwidth is 16.6%. Besides, the 1-bit digital RRM can achieve the beam-scanning. This demonstrates that the beam-scanning performance can be achieved by controlling the coding of the RRM. The measurement results are in good agreement with the simulation results, which shows that the design of the 1-bit digital RRM is effective. Compared with the state-of the-art, the 1-bit digital RRM is a better choice in practical implementations owing to the easier bias voltage control. That is to say, the number of the bias voltage only needs to be two to obtain the coding "1" and coding "0". The 1-bit digital RRM is feasible to achieve beam-scanning and it is preferable to implement, especially in large-scale antenna designs due to stable phase states, system simplification and low cost. In the future, the 1-bit or multi-bit digital RRM will play an essential role in many realms with its unique features.

Acknowledgments: This work is supported by National Natural Science Foundation of China under Contract No. 51477126, and supported by Technology Explorer and Innovation Research Project, and Fundamental Research Funds for the Central Universities (K5051202051 and SPSZ021409).

Author Contributions: All authors contributed substantially to the reported work. Long Li was the originator of the idea of this study and revised the paper; Haixia Liu provided the main instructions of experiments; Shuncheng Tian performed the experiments, analyzed the data and wrote the paper.

Conflicts of Interest: The authors declare no conflict of interest.

References

1. Hum, S.V.; Perruisseau-Carrier, J. Reconfigurable reflectarrays and array lenses for dynamic antenna beam control: A review. *IEEE Trans. Antennas Propag.* **2014**, *62*, 183–198. [CrossRef]
2. Huang, J.; Encinar, J.A. *Reflectarray Antennas*; Springer: Singapore, 2016.
3. Hum, S.V.; Okoniewski, M.; Davies, R.J. Modeling and design of electronically tunable reflectarrays. *IEEE Trans. Antennas Propag.* **2007**, *55*, 2200–2210. [CrossRef]
4. Riel, M.; Laurin, J.J. Design of an electronically beam scanning reflectarray using aperture-coupled elements. *IEEE Trans. Antennas Propag.* **2007**, *55*, 1260–1266. [CrossRef]
5. Rodriguez-Zamudio, J.; Martinez-Lopez, J.I.; Rodriguez-Cuevas, J.; Martynyuk, A.E. Reconfigurable reflectarrays based on optimized spiraphase-type elements. *IEEE Trans. Antennas Propag.* **2012**, *60*, 1821–1830. [CrossRef]
6. Kamoda, H.; Iwasaki, T.; Tsumochi, J.; Kuki, T.; Hashimoto, O. 60-GHz electronically reconfigurable large reflectarray using single-bit phase shifters. *IEEE Trans. Antennas Propag.* **2011**, *59*, 2524–2531. [CrossRef]
7. Chaharmir, M.R.; Shaker, J.; Cuhaci, M.; Sebak, A.R. Novel photonically-controlled reflectarray antenna. *IEEE Trans. Antennas Propag.* **2006**, *54*, 1134–1141. [CrossRef]
8. Perez-Palomino, G.; Baine, P.; Dickie, R.; Bain, M.; Encinar, J.; Cahill, R.; Barba, M.; Toso, G. Design and experimental validation of liquid crystal-based reconfigurable reflectarray elements with improved bandwidth in F-band. *IEEE Trans. Antennas Propag.* **2013**, *61*, 1704–1713. [CrossRef]
9. Carrasco, E.; Tamagnone, M.; Perruisseau-Carrier, J. Tunable graphene reflective cells for THz reflectarrays and generalized law of reflection. *Appl. Phys. Lett.* **2013**, *102*, 183–947. [CrossRef]
10. Hu, W.; Cahill, R.; Encinar, J.A.; Dickie, R.; Gamble, H.; Fusco, V. Design and measurement of reconfigurable millimeter wave reflectarray cells with nematic liquid crystal. *IEEE Trans. Antennas Propag.* **2008**, *56*, 3112–3117. [CrossRef]
11. Perruisseau-Carrier, J. Versatile reconfiguration of radiation patterns, frequency and polarization: A discussion on the potential of controllable reflectarrays for software-defined and cognitive radio systems. In *IEEE International Microwave Workshop Series on RF Front-Ends for Software Defined and Cognitive Radio Solutions*; IEEE: Piscataway, NJ, USA, 2010; pp. 1–4.

12. Li, L.; Jun, C.T.; Ji, W.; Liu, S.; Ding, J.; Wan, X.; Yan, B.L.; Jiang, M.; Qiu, C.; Zhang, S. Electromagnetic reprogrammable coding-metasurface holograms. *Nat. Commun.* **2017**, *8*, 197. [CrossRef] [PubMed]

13. Chen, K.; Feng, Y.; Monticone, F.; Zhao, J.; Zhu, B.; Jiang, T.; Zhang, L.; Kim, Y.; Ding, X.; Zhang, A.; et al. A Reconfigurable Active Huygens' Metalens. *Adv. Mater.* **2017**, *29*. [CrossRef]

14. Mailloux, R.J. Phased array antenna handbook. *Syst. Eng. Electr.* **2011**, *28*, 1816–1818.

15. Hansen, R.C. *Phased Array Antennas*, 2nd ed.; John Wiley & Sons Inc.: Hoboken, NJ, USA, 2009.

16. Bhattacharyya, A.K. *Phased Array Antennas: Floquet Analysis, Synthesis, BFNs, and Active Array Systems*; Wiley: Hoboken, NJ, USA, 2006.

applied
sciences

MDPI

Article

Study of Energy Scattering Relation and RCS Reduction Characteristic of Matrix-Type Coding Metasurface

Jia Ji Yang [1], Yong Zhi Cheng [2,*], Dong Qi [1] and Rong Zhou Gong [1,*]

[1] School of Optical and Electronic Information, Huazhong University of Science and Technology, Wuhan 430074, China; yangjiajialnow@163.com (J.J.Y.); qidong@hust.edu.cn (D.Q.)
[2] Engineering Research Center for Metallurgical Automation and Detecting Technology, Wuhan University of Science and Technology, Wuhan 430081, China
* Correspondence: chengyz@wust.edu.cn (Y.Z.C.); rzhgong@hust.edu.cn (R.Z.G.)

Received: 8 July 2018; Accepted: 19 July 2018; Published: 26 July 2018

Abstract: In this paper, we present a design of the linear polarization conversion metasurface (MS) for the broadband radar cross section (RCS) reduction based on split-ring resonator (SRR) structure in microwave region. The corresponding phase gradient can be obtained through the stable phase difference of basic units of polarization conversion MS. The designed polarization conversion MS is applied in coded electromagnetic (EM) matrix by defining two basic units "0" and "1", respectively. Based on the principle of planar array theory, a new random coding method named by matrix-type coding is proposed. Correlative RCS reduction mechanism is discussed and verified, which can be used to explore the RCS reduction characteristic. The simulated linear polarization conversion rate of the designed structure is up to 90% in the frequency range of 6–15 GHz, and the RCS reduction results verify the theoretical assumptions. Two kinds of matrix-type coding MS samples are prepared and measured. The experimental results indicate that the reflectance of MS is less than –10 dB on average under normal incidence in frequency range of 5.8–15.5 GHz. The average RCS reduction is essentially more than 10 dB in frequency range of 5.5–15 GHz and the corresponding relative bandwidth is 92.7%, which reasonably agrees with simulation. In addition, excellent RCS reduction characteristic of the designed MS can also be achieved over a wide incident angle.

Keywords: polarization conversion; matrix-type coding; energy scattering; RCS reduction

---

1. Introduction

As a two-dimensional artificial material, metasurface (MS) is composed of sub-wavelength element array [1–4], which has been widely applied in optoelectronics devices [5–10], such as sensor, detector, etc. Because of its ability to effective manipulation of electromagnetic (EM) waves, MS can be especially applied in radar stealth field [11–14]. In radar stealth field, the radar echo feature signal can be changed to reduce the detection probability of objects [15–17]. The radar cross section (RCS) is an important physical quantity to measure the echo capability of target radar, which attracts much attention in stealth platforms of military applications [18–20]. By designing MS with different sizes and arrangements, effective RCS reduction can be achieved [21–24].

In recent years, as an important branch of MS, phase gradient metasurface (GMS) with low RCS has been paid great interest due to its tremendous potential in military practice [25–28]. This kind of MS merely reflects the incident waves into the backward space rather than transforming EM energy into heat, which lowers the probability of MS being detected by infrared devices [29–32]. GMS can introduce the artificial wave vector at in-plane direction to control the propagation direction of transmitted and reflected wave beams [33–36]. The polarization conversion characteristic is used to achieve stable phase difference in a broadband frequency range, which can be applied in the design of GMS. By randomly arranging the basic units in GMS, the incident wave is irregularly reflected back to free space, the scattering energy at each directional beam is small. This designed GMS can be used to achieve the RCS reduction characteristic [37–42]. More recently, a GMS based on cruciform structure is proposed [33], which can be applied to the RCS reduction characteristic at low frequency ranges, but the relative bandwidth of designed GMS is narrow. Then, a checkerboard MS based on fishbone-shaped is proposed [41], which can achieve a broadband RCS reduction from 6 to 18 GHz. However, the magnitude of the RCS reduction is about 5 dB, and the design scheme is relatively complex. After this, a new concept of checkboard MS is proposed [42], which can achieve a 10 dB RCS reduction in the frequency range of 9.9–19.4 GHz, but the relative bandwidth is not enough. Further, the polarization-independent MS structure is proposed for RCS reduction [22], which can achieve an ultra-broadband 10 dB RCS reduction characteristic in the range of 17–42 GHz, but the RCS cannot be reduced at the lower frequency, which is still a challenge for practical application. Therefore, it is meaningful to explore the new coding ways of MS, which can manipulate transmitted and reflected EM waves at will to achieve a high relative bandwidth at the lower frequency.

In this paper, the matrix-type random coding theory and the RCS reduction analysis were presented, which revealed a simple and effective method to achieve wideband RCS reduction. Then, six kinds of MSs based on split-ring resonator (SRR) structure were designed to explore the RCS reduction characteristic. Compared with previous works [21–24,33–42], our design has some advantages: Firstly, new random coding method with novel mechanism; Secondly, the structure of basic unit is simple, which can achieve the characteristics of 180° cross-polarization phase differences by simply rotating the metal cut-wire structure; Thirdly, our design presents the RCS reduction at lower frequency with high relative bandwidth compared with the works published before. Such a simple and effective design may provide some potential applications in the field of stealth.

2. Design of Matrix-Type Coding Metasurface

2.1. Matrix-Type Random Coding Theory and RCS Reduction Analysis

Based on the principle of reflective antenna array, a series of random coding MSs with different combinations are designed. The incident wave can be diffusely reflected through the design of the array MS [43–45]. According to "energy conservation law", an effective RCS reduction under normal incidence can be achieved by enhancing the EM energy scattering at other direction.

Assuming the surface is composed of A × B array elements, each array element consists of two basic elements: "0" and "1". The concept of RCS reduction can be explained by the principle of planar array theory [46]. Under normal incidence, the array factor of MS can be expressed as:

$$AF = \sum_{a=1}^{A} \sum_{b=1}^{B} e^{j[(a-1/2)(kd\sin\theta\cos\varphi)+(b-1/2)(kd\sin\theta\sin\varphi)+\phi(a,b)]} \tag{1}$$

where θ and φ are the angles of elevation and azimuth, $k = 2\pi/\lambda$, d is the distance between the basic elements, and $\phi(a,b)$ is the initial phase of the lattice. In our design, the characteristic of cross-polarization reflection phase difference of 180° can be achieved by rotating the basic unit simply, which avoids the complex design of structure size and arranges the MS in a simple and effective way.

A matrix-type random coding way is proposed to design 2-bit coding MS, and the coding flowchart is shown in Figure 1. The basic units of "0" and "1" are placed in the matrix with a fixed ratio. In the case of "0" and "1" with same number, the co-polarization reflection phase difference is 0° and the cross-polarization reflection phase difference is 180°. Therefore, the cross-polarization component of the scattered field can be effectively canceled; leading to a better RCS reduction effect than the traditional random coding MS (the probability of "0" or "1" is 50% in each matrix unit). If the number of units "0" and "1" is not consistent, the cross-polarization component of the scattered field cannot be effectively canceled, leading to a suppression of RCS reduction. A simulation is presented to verify this assumption.

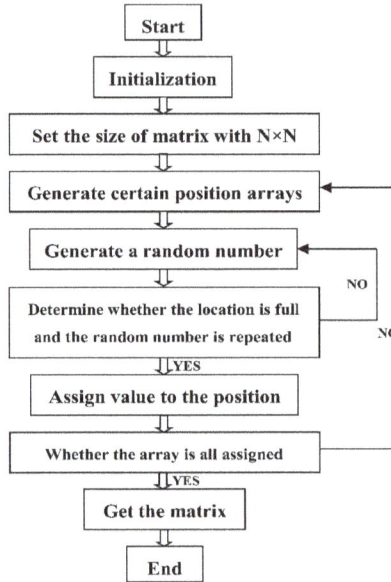

Figure 1. Matrix-type random coding flowchart.

Matrix-type random coding mode M_{random} for Matlab calculation can be expressed as the following functional form:

$$M_{random} = round\left\{ mod\left[\sum_{i=1}^{N} kron\left(rand(2^i), ones(2^{N-i}) \right) \right], 1 \right\} \tag{2}$$

where $rand(2^i)$ and $ones(2^{N-i})$ represent the numbers of block matrix $2^i \times 2^i$, each block matrix composed of $2^{N-i} \times 2^{N-i}$ is the same random number. The "*kron*" is matrix multiplication, where the Kronecker product A, B represents the larger matrix formed by the product of all the elements of matrix A and B. The open interval range of random number is (0, 1). The step to obtain the random coding pattern of the MS is as follows: Firstly, the number of different random matrices in the range of (1, N) is counted. Secondly, the number of patterns is calculated. Finally, the random numbers (0 and 1) are added to the discrete binary codes "0" and "1", and a random coding matrix is obtained through the operation flow chart. Based on the results of simulation and optimization, we choose $i = 6$ and $N - i = 5$ to satisfy the preparation and measurement of sample. Therefore, the designed MS consists of a 6×6 supercell, with each supercell consisting of 5×5 basic units.

The general RCS expression of the scattering surface can be expressed by [46]:

$$\sigma = \lim_{r \to \infty} 4\pi r^2 \frac{S_s}{S_i} = \lim_{r \to \infty} 4\pi r^2 \left| \frac{E_s}{E_i} \right|^2 = \lim_{r \to \infty} 4\pi r^2 \left| \frac{H_s}{H_i} \right|^2 \tag{3}$$

where S_i and S_s are the energy density of incidence and scattering, respectively; $|E_i|$ and $|E_s|$ are the amplitudes of incident and scattered electric fields, respectively; and $|H_i|$ and $|H_s|$ are the amplitudes of the incident and scattered magnetic fields, respectively. The general RCS can also be expressed in the form of *dBsm*:

$$\sigma(dBsm) = 10\lg\left|\sigma(m^2)\right| = 10\lg\left[\lim_{r\to\infty}4\pi r^2\left|\frac{E_s}{E_i}\right|^2\right] \tag{4}$$

In practical military applications, the incident wave can be equivalent to the plane wave because of the transmitting and receiving sources are far from the target. Thus, σ and r are independent of one another. The RCS reduction compared to perfect electric conductor is represented by:

$$\sigma(dBsm)_r = 10\lg\left[\frac{\lim\limits_{r\to\infty}4\pi r^2\left|\frac{E_s}{E_i}\right|^2}{\lim\limits_{r\to\infty}4\pi r^2(1)^2}\right] = 10\lg\left[\left|\frac{E_s}{E_i}\right|^2\right] \tag{5}$$

For a matrix-type random coding MS, each kind of basic unit occupy the half area of total surface in the case of "0" and "1" with same number. The total reflection coefficient can be approximated as the average reflection coefficient of both elements. The RCS reduction can be approximately expressed as [46]:

$$\sigma(dBsm)_{r(1)} = 10\lg\left[\frac{A_0e^{j\vartheta_0} + A_1e^{j\vartheta_1}}{2}\right]^2 \tag{6}$$

where A_0 and A_1 are the reflection coefficient amplitudes of basic units "0" and "1", respectively. ϑ_0 and ϑ_1 are the reflection phases of two basic units. The ratio of "0" and "1" basic units is introduced, which is defined as $\alpha = m_0/m_1$, where m_0 and m_1 are the number of "0" and "1" units, respectively. Further, we introduce α into the formula to express the RCS reduction characteristic with different ratio. However, it can only be used as a qualitative comparison, not a quantitative representation of RCS value. The RCS reduction of 1-bit coding MS under normal incidence can be approximated as:

$$\sigma(dBsm)_{r(\alpha)} = 10\lg\left[\frac{\alpha A_0e^{j\vartheta_0} + A_1e^{j\vartheta_1}}{2}\right]^2 \tag{7}$$

As shown in Equation (7), once basic units of "0" and "1" are selected, the reflection coefficient magnitude and phase can be determined. Therefore, if the ratio α is defined as a constant, the RCS reduction of MS will be fixed. In other words, for fixed α, the magnitudes of RCS reduction with different coding sequences are basically consistent under normal incidence. The ratio, scattered magnitude and phase play important roles in RCS reduction. This assumption is verified by furthering simulation and experiment.

2.2. Matrix-Type Random Coding Metasurface Arrangement

As shown in Figure 2a, the basic unit based on SRR structure is designed, which is set as "0" unit. The SRR structure is rotated counterclockwise by 90° along the wave propagation direction (shown in Figure 2b), which is set as "1" unit. The whole basic unit is divided into three functional layers, where the period is $p = 10$ mm. The upper layer is copper film with SRR structure, which possesses a symmetric axis along 45° with respect to x or y direction. The length of outer radius is $r = 4.1$ mm, the ring width is $s = 0.2$ mm, and the split width of ring structure is $w = 1$ mm. The thickness of the FR4 substrate is 3.5 mm, with the dielectric constant of 4.3 and the loss tangent angle of 0.025. The back layer copper film and the upper layer copper film have the same thickness of 0.035 mm.

Figure 2c presents the co-polarization (r_{xx} and r_{yy}) and cross-polarization (r_{yx} and r_{xy}) reflection coefficients; the high efficient and broadband polarization conversion features can be achieved in a broadband frequency range. The cross-polarization reflection coefficients (r_{yx} and r_{xy}) are greater than 0.8, while the co-polarization reflection coefficients (r_{xx} and r_{yy}) are substantially less than 0.35 in the frequency range of 6–15 GHz. The polarization conversion capability is defined as follows [47]: $PCR_x = |r_{yx}|^2/(|r_{yx}|^2 + |r_{xx}|^2)$ and $PCR_y = |r_{xy}|^2/(|r_{xy}|^2 + |r_{yy}|^2)$. As shown in

Figure 2d, the linear polarization conversion ratio of the x- and y-polarized waves is as high as 85% and reached 99% at resonance frequencies.

Figure 2e,f shows the cross-polarization phase and phase difference of "0" and "1" basic units, respectively. It can be observed that the phase of the "0" and "1" basic unit is different, although the magnitude of the reflection coefficients of co- and cross-polarization is the same. In addition, the phase gradient of designed MS is 180° in the frequency range of 5–16 GHz. Therefore, we use the characteristic of cross-polarization phase difference ±180° to design the matrix-type random coding MS, which can be applied to achieve the broadband RCS reduction.

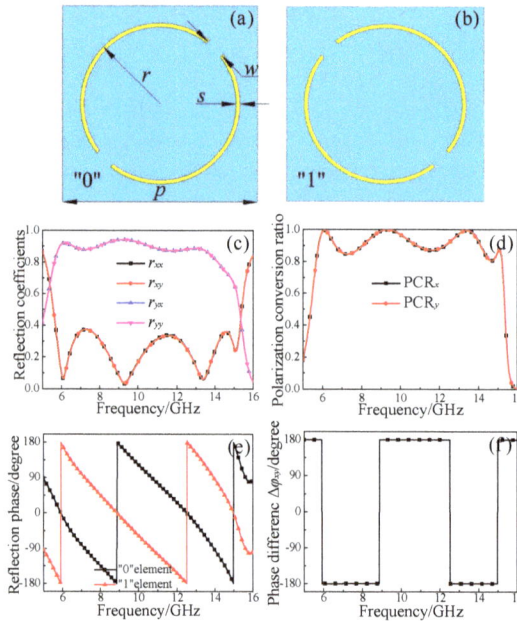

Figure 2. (**a**) Basic unit "0"; (**b**) basic unit "1"; (**c**) the reflection coefficients of "0" and "1"; (**d**) the linear polarization conversion ratio for the normal incident x- and y-polarized wave; (**e**) the cross-polarization phase of "0" and "1"; and (**f**) the cross-polarization phase difference of "0" and "1".

To verify the proposed hypothesis and explore the RCS reduction characteristic of matrix-type random coding MSs, six kinds of arrangements with different ratio of "0" and "1" are designed, in which m_0 and m_1 are the number of "0" and "1" basic units. As for the two basic units, the reflection coefficients and polarization conversion rates are consistent, the units "0" and "1" can be interchanged, and the meaning of the expression is the same. Figure 3a,b presents the schematics of coding a and b with the ratio $\alpha = m_0/m_1 = 1/1$. Figure 3c,d presents the schematics of coding c and d with the ratio $\alpha = m_0/m_1 = 5/4$. Meanwhile, the coding e and f with the ratio $\alpha = m_0/m_1 = 2/1$ are shown in Figure 3e,f. By applying the matrix-type random coding, the direction of energy scattering can be changed to form a diffuse reflection for the incident EM waves; it is possible to achieve the high efficient RCS reduction characteristic under normal incidence.

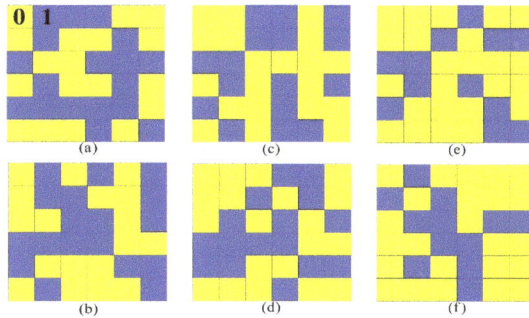

Figure 3. Arrangements of six matrix-type random coding MS: (**a,b**) coding *a* and *b* with the ratio $\alpha = m_0/m_1 = 1/1$; (**c,d**) coding *c* and *d* with the ratio $\alpha = m_0/m_1 = 5/4$; and (**e,f**) coding *e* and *f* with the ratio $\alpha = m_0/m_1 = 2/1$.

To meet the periodic boundary conditions required of simulation, the 5×5 basic units are set as a supercell, and a series of coding arrangements are designed to explore the RCS reduction characteristic of MS. Figure 4a–h presents the far-field scattering characteristic diagram of different coding MSs at 9.5 GHz with the area of 300×300 mm^2. Figure 4a presents the scattering characteristic of metal plate with a strong normal scattering capability, which can be used as a reference. Figure 4b shows the upright energy scattering direction of coding 0 or 1, which is the same as the energy scattering of metal plate. Figure 4c–h shows the scattering characteristic of coding *a* to *f*, the incident EM wave energy scattering is diverged to all around, and the scattering capability is relatively weak at single direction. Thus, these results indicate that the matrix-type random coding MSs have good scattering performance.

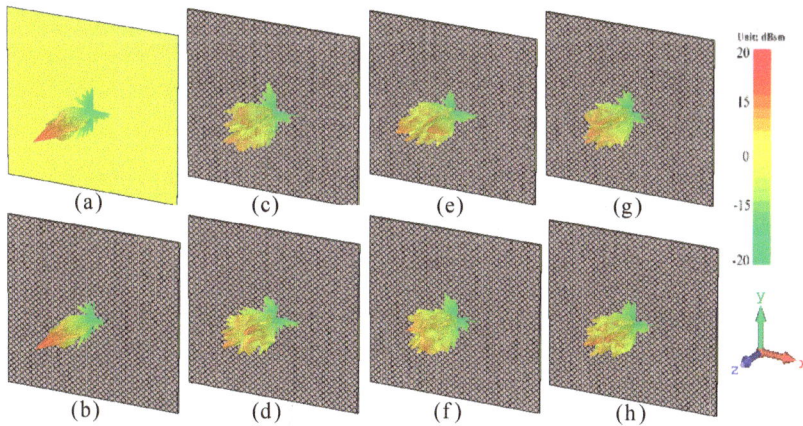

Figure 4. Far-field scattering results of: (**a**) metal plate; (**b**) coding 0 or 1, and matrix-type random coding MS; (**c**) coding *a*; (**d**) coding *b*; (**e**) coding *c*; (**f**) coding *d*; (**g**) coding *e*; and (**h**) coding *f* at 9.5 GHz.

3. Simulation and Experiment

3.1. Simulation and Analysis of Matrix-Type Coding Metasurfaces

The frequency domain solver in EM simulation software of CST MICROWAVE STUDIO is used to perform the numerical simulation. As shown in Figure 5a, the numerical results of coding *a* to *f* are depicted to explore the RCS reduction characteristics of different ratio combinations. The coding 0 presents the RCS reduction of coding 0 or 1, the numerical value is essentially zero in the frequency range of 5–16 GHz, which means the single coding MS cannot reduce the RCS. The coding *a* and *b* are the MSs with the ratio $\alpha = m_0/m_1 = 1/1$; the coding *c* and *d* present the ratio $\alpha = m_0/m_1 = 5/4$; and the coding *e* and *f* present the ratio $\alpha = m_0/m_1 = 2/1$. It can be seen clearly that the RCS reduction curves of different coding sequences with fixed ratio α are basically consistent. The RCS reduction will increase with decrease of the numerical value of α in the whole interested frequency range. The optimal RCS reduction result is presented at $\alpha = m_0/m_1 = 1/1$; these results verify the above theoretical assumptions.

As shown in Figure 5b, the RCS reduction curves of coding *a* are basically consistent at *x*- and *y*-polarized wave incidence, which indicates a polarization-insensitive property of the proposed MSs. In addition, the RCS reduction of coding *a* is greater than 8 dB in the whole frequency range of 5.5–15 GHz, and the RCS reduction reaches a maximum of 21 dB at 9.5 GHz.

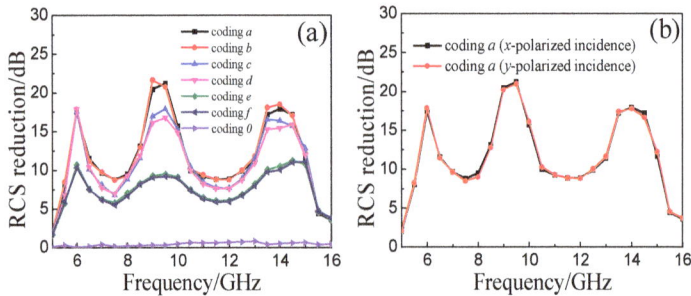

Figure 5. (**a**) RCS reduction of matrix-type random coding MS with different ratio combinations of "0" and "1" units; and (**b**) RCS reduction of coding *a* at *x*- and *y*-polarized wave incidence.

To further discuss the energy scattering characteristic of matrix-type random coding MS, the scattering patterns of coding *a* under normal incidence are studied, as shown in Figure 6a–f. Here is a comparison of coding *a* and metal plate with the same size 300×300 mm^2 at 5, 8, 9.5, 10, 13, and 16 GHz. According to "energy conservation law", the main lobe energy can be suppressed by enhancing the scattering EM energy of side lobe, so an effective RCS reduction can be achieved under normal incidence. The metal plate has a strong main lobe in whole interested frequency range. As shown in Figure 6a,f, the MS has almost no inhibitory effect on main lobe at 5 and 16 GHz. Figure 6b,e shows that the main lobe energy of MS has a certain suppression compared with the metal plate at 8 and 13 GHz. The scattering EM energy is scattered to all around, as shown in Figure 6c,d, which indicates the MS has a significant inhibitory effect on main lobe at 9.5 GHz and 10 GHz, respectively. Generally, the closer it is to the center frequency of the basic unit, the better the effect of reducing RCS can be achieved. Thus, the matrix-type random coding MS allows a wideband and high efficient RCS reduction by adjusting the scattered field simply compared with works proposed before [15,16].

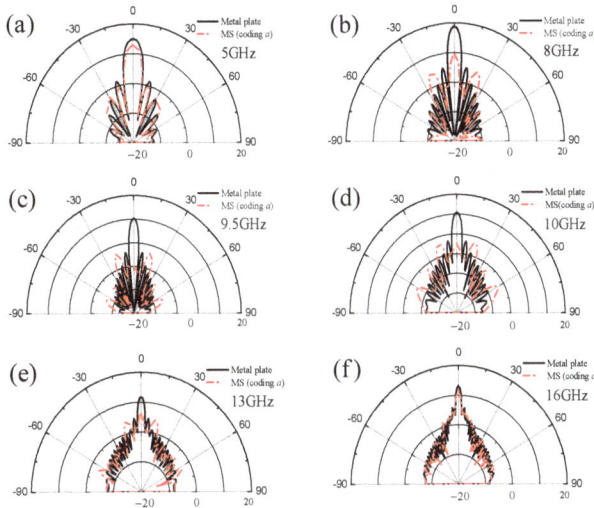

Figure 6. 2D scattering patterns of the coding *a* and metal plate in the *xoz*-plane at: (**a**) 5 GHz; (**b**) 8 GHz; (**c**) 9.5 GHz; (**d**) 10 GHz; (**e**) 13 GHz; and (**f**) 16 GHz.

3.2. Measurement and Analysis of Matrix-Type Coding Metasurface

To further verify the theoretical assumptions and numerical simulation, as shown in Figure 7a,b, two MS samples of coding *a* and *b* were fabricated and measured. The area of MS is 300 × 300 mm², and the thickness of overall design is 3.57 mm. Each sample consists of 6 × 6 supercells and each supercell consists of 5 × 5 basic units of "0" or "1". As shown in Figure 7c, the two samples were measured in the EM anechoic chamber, the transmitting and receiving horns were fixed on the same height level in front of the foam tower. The horn antenna connected to Agilent Technologies N5244A Vector Analyzer was used to measure the RCS of samples. Firstly, the empty darkroom was calibrated before measuring the MS sample. Secondly, the metal ball was placed on the foam tower for calibration as a reference. Thirdly, the MS sample and metal plate were placed on the foam tower for testing to get the RCS value. In measurement, the area of the MS sample was the same as the metal plate. Finally, the RCS reduction of MS sample plate could be obtained by comparing the RCS value of the MS with the metal plate.

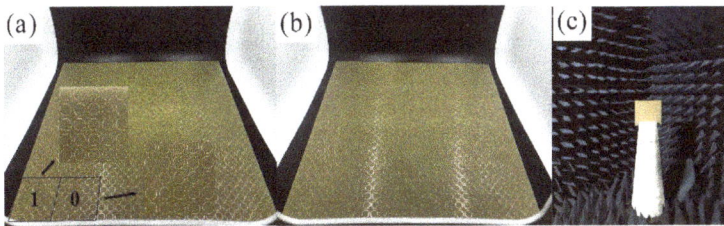

Figure 7. The matrix-type random coding MS templates: (**a**) coding *a*; (**b**) coding *b*; and (**c**) the measurement setup at microwave anechoic chamber.

The reflectance of simulation and experiment at *x*- and *y*-polarized wave incidence are shown in Figure 8a,b. The simulated results are slightly different from the measured ones, which is mainly due to the error occurring in preparation, such as the flatness and the thickness of samples. On the whole, the simulated and measured curves of coding *a* and *b* are consistent well in the entire frequency range of 2–18 GHz, which reveals the polarization-insensitivity of matrix-type random coding MS. The reflectance of coding *a* and *b* are less than −10 dB on average in the frequency range of 5.8–15.5 GHz under normal incidence, which presents an excellent broadband RCS reduction characteristic compared with other works [6–8].

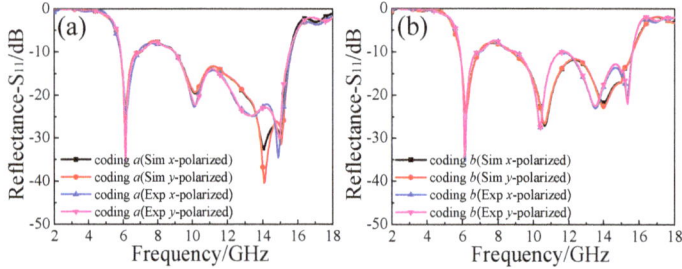

Figure 8. The simulated and measured reflectances of the MS samples at *x*- and *y*-polarized wave incidence: (**a**) coding *a*; and (**b**) coding *b*.

Figure 9a,b shows the RCS reduction of the MS samples under oblique incidence. The RCS reduction is greater than 10 dB in the frequency range of 5.5–15 GHz under normal incidence. For a small oblique incidence of 10°, the average magnitude of RCS reduction is basically consistent with the one of 0°. With the incident angle increasing to 20° and 30°, the RCS reduction effect is suppressed obviously. However, it still presents a more than 5 dB of RCS reduction in a broadband frequency range. A comparison of performance (Table 1) shows that our design has a superior performance in bandwidth and magnitude of RCS reductions compared with the previous reported works [22,33,41,42]. In other words, a considerable RCS reduction of matrix-type random coding MS can be achieved in oblique incident case, which further verifies the excellent RCS reduction characteristic within a wide range of incident angles compared with previous works [17,18].

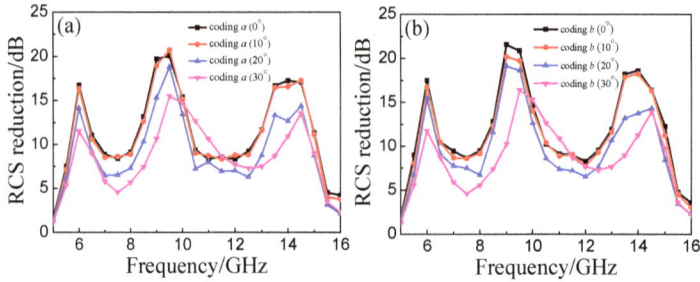

Figure 9. Measured RCS reduction of samples under normal incidence and oblique incidence of 10°, 20°, and 30°: (**a**) coding *a*; and (**b**) coding *b*.

Table 1. Comparison of the designed metasurface with similar works presented in the literature.

Ref.	O. BW. (GHz)	R. BW. (%)	RCS R. (dB)
[33]	3.1–3.4	9.2	10
[41]	6–18	100	5
[42]	9.9–19.4	64.8	10
[22]	17–42	84.7	10
This work	5.5–15	92.7	10

O. BW.: Operation bandwidth; R. BW.: Relative bandwidth; RCS R.: RCS reduction.

4. Conclusions

In this study, a series of matrix-type random coding MSs is designed to explore the RCS reduction characteristic. The coding method of MS with different ratio of "0" and "1" units is proposed, and then, the RCS reduction mechanism of different arrangements is discussed theoretically. To analyze the energy scattering characteristic of random coding sequences, the designed MSs were simulated to get the RCS reduction curves and the scattering patterns under normal incidence. For MS samples of coding *a* and *b*, the reflectance was less than -10 dB on average under *x*- and *y*-polarized wave incidence in the frequency range of 5.8–15.5 GHz, the average RCS reduction is basically larger than 10 dB under normal incidence in the frequency range of 5.5–15 GHz. The matrix–type coding MS presents a broadband RCS reduction characteristic. At oblique incidence, an effective RCS reduction characteristic can also be achieved. Compared with previous works [21–24,33–42], our design of the matrix-type random coding MSs has better broadband RCS reduction and wide-angle incidence tolerance. The designed MS is expected to have potential applications in the field of stealth.

Author Contributions: J.J.Y. and Y.Z.C. designed and performed the experiments; D.Q. and R.Z.G. analyzed the data and contributed reagents/materials/analysis tools; J.J.Y. wrote the paper; and Y.Z.C. and R.Z.G. revised the paper.

Funding: This work was supported by the National Natural Science Foundation of China (U1435209, and 61605147), and the Natural Science Foundation of Hubei China (Grant No. 2017CFB588).

Conflicts of Interest: The authors declare no conflict of interest.

References

1. Pendry, J.B.; Schurig, D.; Smith, D.R. Controlling electromagnetic fields. *Science* **2006**, *312*, 1780–1782. [CrossRef] [PubMed]
2. Martin, F.; Falcone, F.; Bonache, J.; Marques, R. Miniaturized coplanar waveguide stop band filters based on multiple tuned split ring resonators. *IEEE Microw. Wirel. Compos. Lett.* **2003**, *13*, 511–513. [CrossRef]
3. Cui, T.J.; Qi, M.Q.; Wan, X.; Zhao, J.; Cheng, Q. Coding metamaterials, digital metamaterials and programmable metamaterials. *Light Sci. Appl.* **2014**, *3*, 218. [CrossRef]
4. Smith, D.R.; Pendry, J.B.; Wiltshire, M.C.K. Metamaterials and negative refractive index. *Science* **2004**, *305*, 788–792. [CrossRef] [PubMed]
5. Iovine, R.; La Spada, L.; Vegni, L. Modified bow-tie nanoparticles operating in the visible and near infrared frequency regime. *Adv. Nanopart.* **2013**, *2*, 21. [CrossRef]
6. Iovine, R.; La Spada, L.; Vegni, L. Nanoparticle device for biomedical and optoelectronics applications. *COMPEL Int. J. Comput. Math. Electr. Electron.* **2013**, *32*, 1596–1608. [CrossRef]
7. La Spada, L.; Iovine, R.; Tarparelli, R.; Vegni, L. Conical nanoparticles for blood disease detection. *Adv. Nanopart.* **2013**, *2*, 259. [CrossRef]
8. Iovine, R.; La Spada, L.; Vegni, L. Nanoplasmonic sensor for chemical measurements. *Opt. Sens.* **2013**, *8774*, 877411.
9. La Spada, L.; Vegni, L. Near-zero-index wires. *Opt. Express* **2017**, *25*, 23699–23708. [CrossRef] [PubMed]
10. Novoselov, K.S.; Fal'ko, V.I.; Colombo, L.; Gellert, P.R.; Schwab, M.G.; Kim, K. A roadmap for graphene. *Nature* **2012**, *490*, 192. [CrossRef] [PubMed]
11. Wang, P.; Zhang, Y.; Chen, H.; Zhou, Y.; Jin, F.; Fan, H. Broadband radar absorption and mechanical behaviors of bendable over-expanded honeycomb panels. *Compos. Sci. Technol.* **2018**, *162*, 33–48. [CrossRef]
12. Mittal, G.; Pathak, N.P. Design, analysis and characterisation of spoof surface plasmon polaritons based wideband bandpass filter at microwave frequency. *Def. Sci. J.* **2018**, *68*, 300–306. [CrossRef]
13. Zheludev, N.I.; Kivshar, Y.S. From metamaterials to metadevices. *Nat. Mater.* **2012**, *11*, 917. [CrossRef] [PubMed]
14. Kildishev, A.V.; Boltasseva, A.; Shalaev, V.M. Planar photonics with metasurfaces. *Science* **2013**, *339*, 1232009. [CrossRef] [PubMed]

15. Qin, F.; Ding, L.; Zhang, L.; Monticone, F.; Chum, C.C.; Deng, J.; Mei, S.; Li, Y.; Teng, Y.; Hong, M.; et al. Hybrid bilayer plasmonic metasurface efficiently manipulates visible light. *Sci. Adv.* **2016**, *2*, E1501168. [CrossRef] [PubMed]

16. La Spada, L.; Haq, S.; Hao, Y. Modeling and design for electromagnetic surface wave devices. *Radio Sci.* **2017**, *52*, 1049–1057. [CrossRef]

17. Cai, W.; Chettiar, U.K.; Kildishev, A.V.; Shalaev, V.M. Optical cloaking with metamaterials. *Nat. Photonics* **2007**, *1*, 224. [CrossRef]

18. Liu, Y.; Hao, Y.; Li, K.; Gong, S. Radar cross section reduction of a microstrip antenna based on polarization conversion metamaterial. *IEEE Antennas Wirel. Propag. Lett.* **2016**, *15*, 80–83. [CrossRef]

19. Song, Y.C.; Ding, J.; Guo, C.J.; Ren, Y.H.; Zhang, J.K. Radar cross-section reduction based on an iterative fast Fourier transform optimized metasurface. *Mod. Phys. Lett. B* **2016**, *30*, 1650233. [CrossRef]

20. Selvaraju, R.; Jamaluddin, M.H.; Kamarudin, M.R.; Nasir, J.; Hashim, D. Complementary split ring resonator for isolation enhancement in 5G communication antenna array. *Prog. Electromagn. Res.* **2018**, *83*, 217–228. [CrossRef]

21. Su, P.; Zhao, Y.J.; Jia, S.L.; Shi, W.W.; Wang, H.L. An ultra-wideband and polarization-independent metasurface for RCS reduction. *Sci. Rep.* **2016**, *6*, 20387. [CrossRef] [PubMed]

22. Sun, H.Y.; Gu, C.Q.; Chen, X.L.; Li, Z.; Liu, L.L.; Xu, B.Z.; Zhou, Z.C. Broadband and broad-angle polarization-independent metasurface for radar cross section reduction. *Sci. Rep.* **2017**, *7*, 40782. [CrossRef] [PubMed]

23. Chen, K.; Cui, L.; Feng, Y.J.; Zhao, J.M.; Jiang, T.; Zhu, B. Coding metasurface for broadband microwave scattering reduction with optical transparency. *Opt. Express* **2017**, *25*, 5571–5579. [CrossRef] [PubMed]

24. Cheng, Y.Z.; Wu, C.J.; Ge, C.C.; Yang, J.J.; Pei, X.J.; Jia, F.; Gong, R.Z. An ultra-thin dual-band phase-gradient metasurface using hybrid resonant structures for backward RCS reduction. *Appl. Phys. B* **2017**, *123*, 143. [CrossRef]

25. Shelby, R.A.; Smith, D.R.; Schultz, S. Experimental verification of a negative index of refraction. *Science* **2001**, *292*, 77–79. [CrossRef] [PubMed]

26. Li, Y.; Zhang, J.; Qu, S. Wideband radar cross section reduction using two-dimensional phase gradient metasurfaces. *Appl. Phys. Lett.* **2014**, *104*, 221110. [CrossRef]

27. Sui, S.; Ma, H.; Wang, J.; Pang, Y.; Feng, M.; Xu, Z.; Qu, S. Absorptive coding metasurface for further radar cross section reduction. *J. Appl. Phys.* **2018**, *51*, 065603. [CrossRef]

28. Alrasheed, S.; Di Fabrizio, E. Design and simulation of reflect-array metasurfaces in the visible regime. *Appl. Opt.* **2017**, *56*, 3213–3218. [CrossRef] [PubMed]

29. Mitrofanov, O.; Viti, L.; Dardanis, E. Near-field terahertz probes with room-temperature nanodetectors for subwavelength resolution imaging. *Sci. Rep.* **2017**, *7*, 44240. [CrossRef] [PubMed]

30. Politano, A.; Viti, L.; Vitiello, M.S. Optoelectronic devices, plasmonics, and photonics with topological insulators. *APL Mater.* **2017**, *5*, 035504. [CrossRef]

31. Kou, N.; Liu, H.; Li, L. A transplantable frequency selective metasurface for high-order harmonic suppression. *Appl. Sci.* **2017**, *7*, 1240. [CrossRef]

32. Li, Y.; Zhang, J.; Qu, S. Achieving wideband polarization-independent anomalous reflection for linearly polarized waves with dispersionless phase gradient metasurfaces. *J. Phys. D Appl. Phys.* **2014**, *47*, 425103–425109. [CrossRef]

33. Wu, C.J.; Cheng, Y.Z.; Wang, W.Y.; He, B.; Gong, R.Z. Design and radar cross section reduction experimental verification of phase gradient meta-surface based on cruciform structure. *Acta Phys. Sin.* **2015**, *64*, 064102.

34. Li, Y.F.; Wang, J.; Zhang, J.; Qu, S.B.; Pang, Y.Q.; Zheng, L.; Yan, M.B.; Xu, Z.; Zhang, A.X. Ultra-wide-band Microwave composite absorbers based on phase gradient metasurfaces. *Prog. Electromagn. Res.* **2014**, *40*, 9–18. [CrossRef]

35. Cheng, Y.Z.; Cheng, Z.Z.; Mao, X.S.; Gong, R.Z. Ultra-thin multi-band polarization-insensitive microwave metamaterial absorber based on multiple-order responses using a single resonator structure. *Materials* **2017**, *10*, 1241. [CrossRef] [PubMed]

36. Yu, N.F.; Genevet, P.; Kats, M.A.; Aieta, F. Light propagation with phase discontinuities: Generalized laws of reflection and refraction. *Science* **2011**, *334*, 333–337. [CrossRef] [PubMed]

37. Deng, G.S.; Xia, T.Y.; Fang, Y.; Yang, J.; Yin, Z.P. A polarization-dependent frequency-selective metamaterial absorber with multiple absorption peaks. *Appl. Sci.* **2017**, *7*, 580. [CrossRef]

38. Huang, C.; Pan, W.B.; Ma, X.L.; Luo, X.G. Wideband radar cross-section reduction of a stacked patch array antenna using metasurface. *IEEE Antennas Wirel. Propag. Lett.* **2015**, *14*, 1369–1372. [CrossRef]

39. Tian, S.; Liu, H.; Li, L. Design of 1-bit digital reconfigurable reflective metasurface for beam-scanning. *Appl. Sci.* **2017**, *7*, 882. [CrossRef]

40. Yang, J.J.; Cheng, Y.Z.; Ge, C.C.; Gong, R.Z. Broadband polarization conversion metasurface based on metal cut-wire structure for radar cross section reduction. *Materials* **2018**, *11*, 626. [CrossRef] [PubMed]

41. Liu, Y.; Li, K.; Jia, Y.; Hao, J.; Gong, S.; Guo, Y.J. Wideband RCS reduction of a slot array antenna using polarization conversion metasurfaces. *IEEE Trans. Antennas Propag.* **2016**, *64*, 326–331. [CrossRef]

42. Zheng, Q.; Li, Y.; Zhang, J.; Ma, H.; Wang, J.; Pang, Y.; Han, Y.; Sui, S.; Shen, Y.; Chen, H.; et al. Wideband, wide-angle coding phase gradient metasurfaces based on Pancharatnam-Berry phase. *Sci. Rep.* **2017**, *7*, 43543. [CrossRef]

43. Iovine, R.; La Spada, L.; Tarparelli, R.; Vegni, L. Spectral green's function for SPR meta-structures. *Mater. Sci. Forum* **2014**, *792*, 110–114. [CrossRef]

44. Padooru, Y.R.; Yakovlev, A.B.; Kaipa, C.S.R.; Hanson, G.W.; Medina, F.; Mesa, F.; Glisson, A.W. New absorbing boundary conditions and analytical model for multilayered mushroom-type metamaterials: Applications to wideband absorbers. *IEEE Trans. Antennas Propag.* **2012**, *60*, 5727–5742. [CrossRef]

45. La Spada, L.; Vegni, L. Electromagnetic nanoparticles for sensing and medical diagnostic applications. *Materials* **2018**, *11*, 603. [CrossRef] [PubMed]

46. Balanis, C.A. *Antenna Theory: Analysis and Design*, 3rd ed.; Wiley: New York, NY, USA, 2005.

47. Zhao, J.C.; Cheng, Y.Z.; Cheng, Z.Z. Design of a photo-excited switchable broadband reflective linear polarization conversion metasurface for terahertz waves. *IEEE Photonics J.* **2018**, *10*, 1–10. [CrossRef]

applied
sciences

MDPI

Article

Characteristic Analysis of Compact Spectrometer Based on Off-Axis Meta-Lens

Yi Zhou, Rui Chen and Yungui Ma *

Centre for Optical and Electromagnetic Research, State Key Lab of Modern Optical Instrumentation, Zhejiang University, Hangzhou 310058, China; zhouyi2016@zju.edu.cn (Y.Z.); chen_rui@zju.edu.cn (R.C.)
* Correspondence: yungui@zju.edu.cn; Tel.: +86-571-88206514-209

Received: 18 January 2018; Accepted: 23 February 2018; Published: 26 February 2018

Abstract: Ultra-compact spectrometers with high-resolution and/or broadband features have long been pursued for their wide application prospects. The off-axis meta-lens, a new species of planar optical instruments, provides a unique and feasible way to realize these goals. Here we give a detailed investigation of the influences of structural parameters of meta-lens-based spectrometers on the effective spectral range and the spectral resolution using both wave optics and geometrical optics methods. Aimed for different usages, two types of meta-lens based spectrometers are numerically proposed: one is a wideband spectrometer working at 800–1800 nm wavelengths with the spectral resolution of 2–5 nm and the other is a narrowband one working at the 780–920 nm band but with a much higher spectral resolution of 0.15–0.6 nm. The tolerance for fabrication errors is also discussed in the end. These provides a prominent way to design and integrate planar film-based spectrometers for various instrumental applications.

Keywords: meta-lens; metasurface; spectrometer; dispersion; off-axis optical system

1. Introduction

Recent progresses on micro/nano-photonics have become a turning point for the transformation of traditional optical systems [1]. Metasurfaces, ultrathin planar structures patterned by monolithic subwavelength elements on surfaces, provide a new way to transform the incident wavefront to the desired forms by controlling the amplitude, phase and polarization of electromagnetic fields [2–6]. Compared with their traditional counterparts, metasurfaces have emerged as unparalleled devices to realize miniaturized, on-chip integrated and multifunctional optical systems [7–10]. Tremendous breakthroughs in the field of metasurface have given rise to expansive applications ranging from planar imaging lenses [8,9,11–14], holograms [15–19], spectrometers [20,21], nonlinear devices [22,23] and even to invisible cloaks [24–26], etc.

Spectroscopy, a fundamental scientific instrument which has developed for hundreds of years since Newton elaborated his discover that the sunlight is made up of a mixture of colors in 1672 [27], is of significant importance to physics [28–30], astronomy [31–33] and chemistry [34–36]. As pivotal components in spectroscopy, traditional spectrometers usually employ isolated dispersive elements (mostly gratings or prisms) and focusing lenses, which makes the whole system bulky, complicated and unsuitable for integration [21]. To produce miniature and compact spectrometers, researchers have explored the possibilities of taking the advantages of metasurfaces such as its flexibility to design arbitrary phase response or convenience to be integrated to other optical system to create multifunctional meta-lens-based spectrometers [20,21,37–39]. Spectrometers based on off-axis meta-lenses was first explored by Khorasaninejad et al. [20], which unambiguously shows the potential of the technique, i.e., meta-lenses, as an alternative way to build spectrometers. However, a more detailed parametric analysis is required to fully understand the features of the two-dimensional device, especially clarifying the performance dependencies (spectral range and spectral resolution) on the system's structural parameters (lens aperture, focal length, off-axis angle and orientation angle of the detector plane). These information is crucial for the device design and the dispersive characteristics of off-axis meta-lenses need to be further illustrated.

In this paper, we carry out a comprehensive characteristic analysis of off-axis meta-lens based compact spectrometers using both rigorous wave theory and geometrical ray optics. The dispersive properties of the off-axis meta-lens provide the technique foundation for spectroscopy. We first show that by optimizing the orientation angle of the output (detector) plane, the effective spectral range can be enhanced about three times than that of the common configuration where the output plane is placed perpendicular to the optical axis. From the numerical diffraction calculations compared with the analytical geometrical optics analysis, we then discuss the full picture about the structural dependence of the intrinsic performance of the device. It is found that to design a high-resolution spectrometer, the aperture diameter and off-axis angle of the meta-lens should be enlarged as much as possible. On the contrary, a broadband spectrometer can be designed at certain sacrifice of spectral resolution by reducing the lens's aperture and off-axis angle and increasing the focal length. Based on these features, we give two concrete device designs: one could cover the 800–1800 nm band with the spectral resolution of 2–5 nm and the other covering the 780–920 nm band with a finer spectral resolution of 0.15–0.6 nm. The current research provides a new perspective to design meta-lens-based spectrometers.

2. Design Principle of Spectrometer Based on Off-Axis Meta-Lens

A schematic diagram of the meta-lens-based spectrometer is shown in Figure 1. For practice, a collimation light circuit is basically required for the proposed device (not shown in Figure 1). For the off-axis meta-lens which will focus the incident normal plane wave at a deflected angle α, the phase distribution φ_d imparted by it follows

$$\varphi_d(x, y) = 2\pi - \frac{2\pi}{\lambda_d} \left(\sqrt{(x - f \sin \alpha)^2 + y^2 + (f \cos \alpha)^2} - f \right), \tag{1}$$

where λ_d is the design wavelength (corresponding to λ_2 in Figure 1) and f is the focal length.

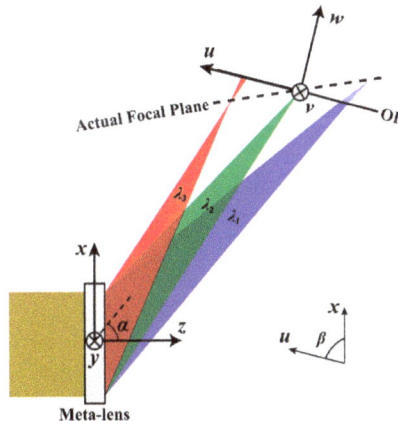

Figure 1. Schematic illustration showing the configuration of the spectrometer based on an off-axis meta-lens. The design wavelength is λ_2 and the off-axis angle is α. The actual focal plane denoted by the dashed line has a skew orientation angle with the optical axis. The angle between the u-axis and the x-axis is defined as the orientation angle of the output plane (OP) β. The inset shows the definition of its sign. In addition to the global coordinate system (x, y, z) based on the meta-lens, a local coordinate system (u, v, w) based on the OP is adopted to simplify the deduction.

Two kinds of chromatics dispersion are introduced according to this configuration—the longitudinal displacement along the optical axis (the line from the center of the meta-lens to the focal point of the design wavelength) and lateral displacement perpendicular to the optical axis [21,40]. It can be predicted that the actual focal plane for light with different wavelengths (λ_1, λ_2 and λ_3 indicated in Figure 1 as an illustration) has a skew orientation angle with the optical axis. The output plane (OP) where the detector is placed does not always coincide with the actual focal plane in practice. A local coordinate system (u, v, w) centered at the focal point of the design wavelength ($f \sin\alpha, 0, f \cos\alpha$) is introduced at a relation with (x, y, z) by

$$\begin{cases} x = u \cos\beta + w \sin\beta + f \sin\alpha, \\ y = v, \\ z = -u \sin\beta + w \cos\beta + f \cos\alpha, \end{cases} \quad (2)$$

where β denotes the angle between the u-axis and the x-axis. At the designing wavelength, the meta-lens will work as a blazed grating. At other frequencies, diffraction will be the dominant factor to decide the output field. In practice, the meta-lenses are usually constructed using subwavelength elements that have ideally a linear dispersion so that there is a dominant diffraction order for different frequencies in the band and their refractive angles are solely decided by the local period. Therefore, in ray optics, the meta-lens could be represented by a constant and frequency-independent phase profile. This assumption is also valid for wave optics when a meta-lens of broadband response is employed, as we assumed, for example the one consisting of Pancharatnam-Berry phase elements working for circularly polarized waves [20,21]. Then, the input field $U_i(x, y, 0)$ is written as

$$U_i(x, y, 0) = P(x, y) \exp(i\varphi_d(x, y)), \quad (3)$$

where $P(x, y)$ is the pupil function of the meta-lens. It takes nonzero and unity value only inside the meta-lens aperture. Based on Rayleigh-Sommerfeld's diffraction formula, the output field $U_o(u, v, 0)$ is expressed by [41]

$$\begin{cases} U_o(u, v, 0) = \frac{1}{i\lambda} \iint U_i(x, y, 0) \frac{\exp(i2\pi r/\lambda)}{r^2}(-u \sin\beta + f \cos\alpha)dxdy, \\ r = \sqrt{(u \cos\beta + f \sin\alpha - x)^2 + (v - y)^2 + (-u \sin\beta + f \cos\alpha)^2}, \end{cases} \quad (4)$$

where r denotes the distance from the source $(x, y, 0)$ to the observation point $(u, v, 0)$ (local coordinate system).

3. Results and Discussion

3.1. Simulation Results of a Typical Meta-Lens-Based Spectrometer

It is pivotal to analyze the dispersive characteristics of the spectrometer based on the off-axis meta-lens first. According to Equations (1)–(4), here we give the simulation results of the configuration with parameters $f = 20$ mm, aperture diameter $D = 2$ mm, $\alpha = 45°$ and $\beta = 45°$ (the OP is perpendicular to the optical axis) at $\lambda_d = 1550$ nm, as shown in Figure 2. It corresponds to a small numerical aperture NA = 0.035 (see Equation (5) below). The focal line profile along the u-axis as λ varies in 1520–1580 nm is shown in Figure 2a. The dispersion-caused spatial focal spot distribution ensures the technical practicability to develop a meta-lens based spectrometer. One can see the focal line gets widened obviously as wavelength significantly deviates from the design wavelength due to the chromatic aberration and the mismatch of the actual focal plane and the OP. In this spectral range, the displacement Δu is nearly linearly dependent on wavelength, suggesting a linear dispersion $du/d\lambda = 9.95$ μm/nm. A diffraction-limited spatial full width at half-maximum (FWHM) of the focal spot of 19.6 μm (12.6 λ_d) at $\lambda_d = 1550$ nm is observed from Figure 2b. It raises slowly as wavelength deviates from λ_d. We introduce the effective spectral range $\Delta\lambda$ to evaluate the performance of the spectrometer. Within this spectral range, the FWHM of the spot line is less than 1.2 times $\text{FWHM}|_{\lambda=\lambda_d}$. In fact, this criterion is much stringent compared with previous studies about optical spectrometers and the commercial products [21]. For the configuration proposed here, $\Delta\lambda = 1574$ nm $- 1526$ nm $= 48$ nm. The spectral resolution $\delta\lambda$ as a function of wavelength define by the FWHM value of the $U_o(\lambda)$ curve is also shown in Figure 2b. The inset of Figure 2b plots the $U_o(\lambda)$ intensity along the dashed line in Figure 2a at $u = 0$. Over the spectral range 1524–1574 nm, the spectral resolution is within 1.8 nm. Figure 2c plots the field intensity (normalized by the maximum of the field intensity at the design wavelength λ_d) profile along the u-axis at different wavelengths at an interval of 1.5 nm. Different spectral lines are distinguishable in this interval. These results are acquired without considering the actual pixel size of the detector.

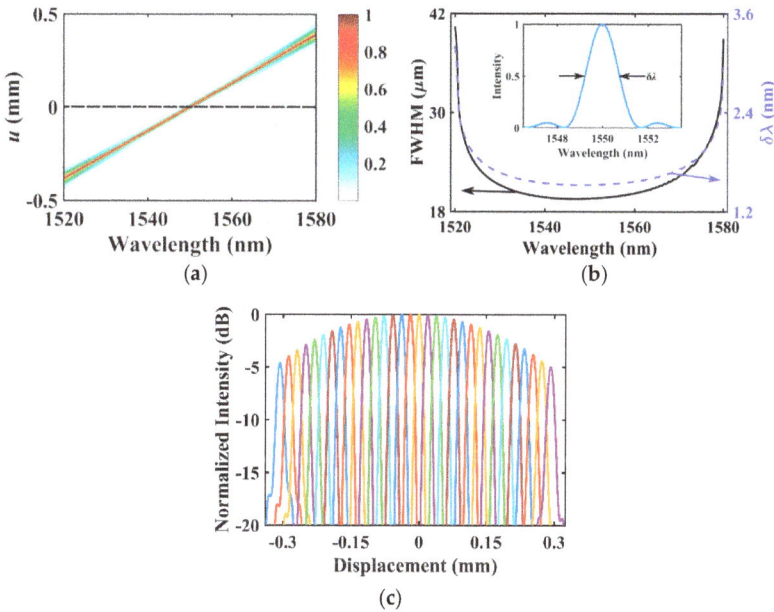

Figure 2. Dispersive characteristics of a meta-lens-based spectrometer: (**a**) The focal line along the u-axis as a function of wavelength for the configuration with $f = 20$ mm, aperture diameter $D = 2$ mm, $\alpha = 45°$ and $\beta = 45°$; (**b**) FWHM and spectral resolution $\delta\lambda$ of the focal line as a function of wavelength. Inset shows the field profile along the dashed line in (**a**); (**c**) Electric field intensity normalized by the maximum of the field intensity at the design wavelength on the OP at different wavelengths (from 1526 to 1574 nm at a constant interval of 1.5 nm).

3.2. Relationship between Structural Parameters and Evaluation Indexes of Spectrometer

As mentioned above, the disorientation of the actual focal plane with the output plane restricts the effective spectral range of the meta-lens-based spectrometer. We simulate the output field of system with different orientation angles of the OP β while other parameters are kept constant as those used in Figure 2 and analyze the effective spectral ranges $\Delta\lambda$ and spectral resolutions $\delta\lambda$ of them (shown in Figure 3a). The lower and upper bounds of the blue band in Figure 3a denote the minimum and maximum spectral resolution, respectively, which roughly have a constant ratio. The spectral range $\Delta\lambda$ reaches the maximum at about 140 nm (almost three times larger than that at $\beta = 45°$ in Figure 2) when β is about $-70°$. According to the above discussions, it is obvious that the effective spectral range will reach maximum when the OP overlaps the actual focal plane because of the decrease of the aberration. On the other hand, when the OP is nearly perpendicular to the actual focal plane, the effective spectral range losses its physical meaning. The spectral resolution $\delta\lambda$ is nearly independent on β and primarily decided by the structural parameters of the meta-lens (they are the aperture diameter, the focal length and the off-axis angle). Although the linear dispersion will increase when the orientation of OP is overlapped with the actual focal plane, the size of the focal spot will also increase proportionally.

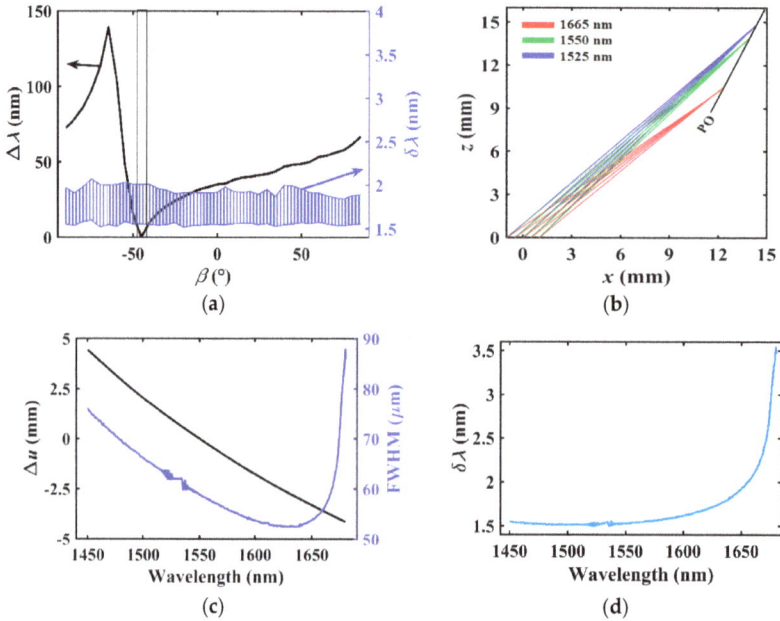

Figure 3. Influences of the orientation angle of the OP β on the dispersive properties of the meta-lens: (a) The effective spectral range $\Delta\lambda$ and spectral resolution $\delta\lambda$ as a function of β. Arrows ascribe the curves to the left or right y-axis. The data for β close to $-45°$ (indicated by the gray box) has low accuracy because the OP is parallel to the optical axis at this specific case; (b) Ray tracing calculations for the meta-lens at wavelengths of 1525, 1550 and 1665 nm, respectively, indicating an optimal orientation angle $\beta = -70°$; (c) Displacement Δu and FWHM of the focal line as a function of wavelength at $\beta = -70°$; (d) Spectral resolution $\delta\lambda$ as a function of wavelength at $\beta = -70°$.

Based on the above analysis, hereafter we go a step further to study the optimal value of β according to the ray tracing method. Adoption of the method is reasonable because the geometrical optics approximation is available for our low NA design. A commercial software Zemax 2005 (Zemax LLC, Kirkland, WA, USA, 2005) was used to trace the light path at three wavelengths (1525 nm, 1550 nm, 1665 nm), as shown in Figure 3b. We adopt the Zernike standard phase surface deduced from Equation (1) [42]. We choose 1525 and 1665 nm as simulation wavelengths because they are the lower and upper bounds of the effective spectral range. The ray tracing results suggest that the optimal orientation angle is $\beta = -70°$, which agrees exactly with the prediction of wave analysis. The displacement Δu and FWHM of the focal line as a function of the operation wavelength at $\beta = -70°$ are shown in Figure 3c. The position of the minimum FWHM of the focal line red-shifts to the wavelength 1640 nm due to the extremely oblique orientation angle relative to the optical axis. The displacement of the focal line is not linearly dependent on wavelength anymore. Figure 3d gives the spectral resolution $\delta\lambda$ as a function of wavelength at $\beta = -70°$. The spectral resolution below 1.6 nm is obtained over a wide spectral range from 1450 to 1600 nm. Although the effective spectral range will be greatly enhanced by adjusting the orientation angle of the OP, the actual working efficiency of the device may degrade due to the reduced photon absorption efficiency of detector at grazing incident angles. In principle, this problem may be overcome by patterning the detector with another metasurface. In the following, we will restrict the discussions by using the condition $\beta = \alpha$ (i.e., the OP is perpendicular to the optical axis). This will not lose the generality to illustrate the features of spectrometers that can be reproduced with different specifications.

There are three factors that affect the performance of the meta-lens-based spectrometer besides β: aperture diameter D, focal length f and off-axis angle α. For the meta-lens, NA is expressed as

$$NA = \sin\left(\frac{1}{2} \tan^{-1} \frac{4F \cos \alpha}{4F^2 - 1}\right),$$ (5)

where the F-number is $F = f/D$. The influences of D, f and α on $\Delta\lambda$ and $\delta\lambda$ are presented in Figure 4. Same as above, the lower and upper bounds of the blue band denote the minimum and maximum spectral resolution, respectively. According to the calculation results, the effective spectral range $\Delta\lambda$ is linearly proportional to the focal length f but the spectral resolution $\delta\lambda$ is nearly independent on f. However, $\Delta\lambda$ and $\delta\lambda$ both show an inverse relation with the aperture diameter D and the off-axis angle α. The configuration with large f or small D and α attains wider effective spectral range due to the smaller spherical aberration [43].

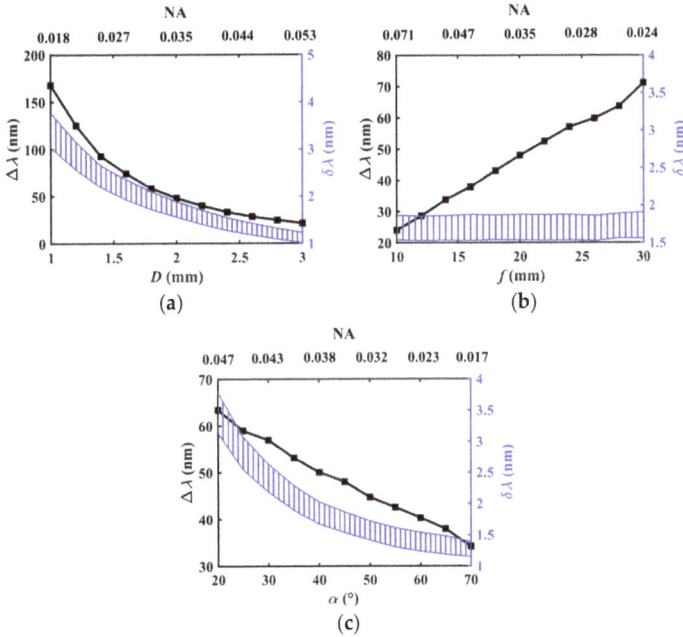

Figure 4. Effective spectral range $\Delta\lambda$ and spectral resolution $\delta\lambda$ as a function of aperture diameter D when $f = 20$ mm and $\alpha = 45°$ (**a**); focal length f when $D = 2$ mm and $\alpha = 45°$ (**b**); off-axis angle α when $D = 2$ mm and $f = 20$ mm.

From the wave theory and the linear dispersion $du/d\lambda$ for $\beta = \alpha$ (see Figure 2b), the spectral resolution can be estimated by [20]

$$\delta\lambda = \frac{d\lambda}{f\{\sin^{-1}[(1 + \frac{d\lambda}{\lambda_d}) \sin \alpha] - \alpha\}} \times \frac{0.5\lambda}{NA}.$$ (6)

For a large *F*-number (*F* > 6 in our case), Equation (5) can be simplified as

$$\text{NA} \approx \frac{1}{2} \tan^{-1} \frac{4F \cos \alpha}{4F^2 - 1} \approx \frac{2F \cos \alpha}{4F^2 - 1}. \tag{7}$$

Then the spectral resolution at the design wavelength $\delta\lambda|_{\lambda=\lambda_d}$ is

$$\delta\lambda|_{\lambda=\lambda_d} = \lim_{d\lambda \to 0} \delta\lambda = \frac{0.5\lambda_d^2}{f \tan \alpha \times \text{NA}} = \frac{\lambda_d^2}{D \sin \alpha}[1 - (\frac{1}{2F})^2] \approx \frac{\lambda_d^2}{D \sin \alpha}. \tag{8}$$

According to Figure 2c, it is reasonable to approximate the minimum spectral resolution to be equal to $\delta\lambda|_{\lambda=\lambda_d}$ for $\beta = \alpha$. Then we conclude that the minimum spectral resolution is inversely proportional to D or the sine of α and nearly independent of f for meta-lens with a large *F*-number. The analytical result according to Equation (8) provides almost the same variation tendency with the numerical results.

3.3. Two Practical Device for Different Applications

Consider the fact that the detector always has limited element sizes and array numbers, there will be a trade-off between the effective spectral range and spectral resolution in designing spectrometer. A high-resolution spectrometer with a wide spectral range is ideal but not realistic for the grating based species. Alternatively, we propose two practical configurations to satisfy the actual requirements: a wideband spectrometer with medium spectral resolution and a narrowband spectrometer with high spectral resolution. In practice, the meta-lens based approach has the advantages to integrate these two or more components into a single package.

The structural parameters for the wideband spectrometer with medium spectral resolution are f = 30 mm, D = 2 mm, α = 15°, β = −89° at λ_d = 1550 nm. Its focal line along the *u*-axis and spectral resolution as a function of wavelength are shown in Figure 5a,b, respectively, indicating a remarkably wide spectral range. This device will work at the band 800–1800 nm with the spectral resolution of 2–5 nm. The dimension of this configuration is estimated as $70 \times 20 \times 10$ mm^3. This configuration is compact and its spectral range and spectral resolution is comparable to the traditional commercial mini-spectrometer that has the similar working parameters, for example, Hamamatsu Photonics C11482GA [44]. The structural parameters for the narrowband spectrometer with high spectral resolution are f = 30 mm, D = 6 mm, α = 45°, β = −65° at λ_d = 850 nm. Its focal line along the *u*-axis and the spectral resolution as a function of wavelength are shown in Figure 5b,c, respectively, indicating a narrow spectral range 780–920 nm. Within this spectral range, the spectral resolution is within 0.15–0.6 nm. Its dimension is approximately $40 \times 30 \times 10$ mm^3. This results are also comparable to the traditional commercial spectrometer such as Hamamatsu Photonics C13054MA [44].

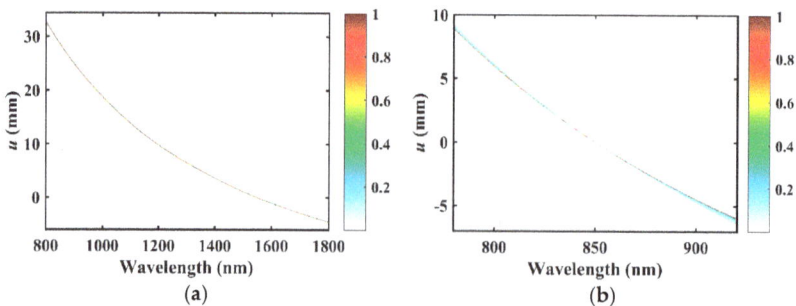

Figure 5. *Cont.*

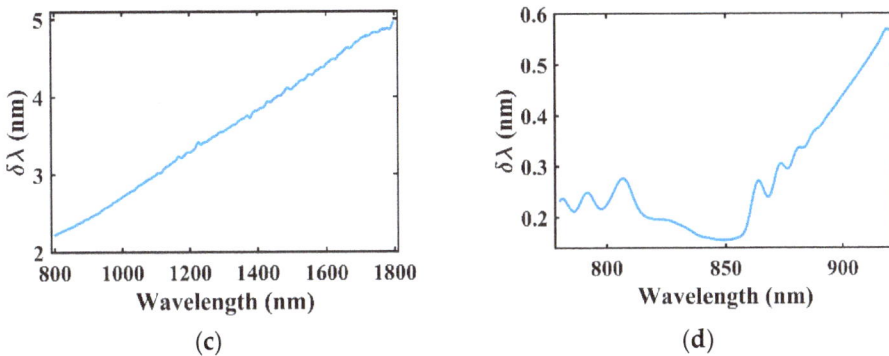

Figure 5. Two practical configurations for the actual application prospects: (**a**) The focal line along the *u*-axis as a function of wavelength for the configuration with f = 30 mm, D = 2 mm, α = 15° and β = −89° at λ_d = 1550 nm; (**b**) The focal line along the *u*-axis as a function of wavelength for the configuration with f = 30 mm, D = 6 mm, α = 45° and β = −65° at λ_d = 850 nm; (**c**) Spectral resolution $\delta\lambda$ as a function of wavelength of the configuration in (**a**); (**d**) Spectral resolution $\delta\lambda$ as a function of wavelength of the configuration in (**b**).

In practice, one can determine the initial parameters of the meta-lens according to Equation (8) to meet the requirements of the spectral resolution. The effective spectral range could be broadened obviously by adjusting the orientation angle of the OP β according to the ray tracing method. These parameters together give the freedoms to design spectrometer for various application requirements.

3.4. Influence of the Fabrication Error Analysis

In this section, we give a numerical analysis how the imperfect implementation may affect the performance of the proposed device. To do so, we have assumed that any imperfection in fabrication may cause the phase distortion of the ideal device from Equation (1).

Figure 6a gives the effective spectral range $\Delta\lambda$, the spectral resolution $\delta\lambda$ and the maximum intensity I_m of the focal spot. The latter two are taken at the design wavelength. In the calculation, a random noise using the built-in function of the MATLAB is applied to the phase profile in Equation (1), whose value obeys a normal distribution $\mathcal{N}(0, \sigma^2)$. Here, σ denotes the standard deviation. It is seen that the maximum intensity I_m decreases when the value σ increases, indicating the weakened focusing efficiency at larger phase distortion. However, the effective spectral range and the spectral resolution at the design wavelength almost remain stable as σ changes because they are solely contributed by the zeroth diffraction order. These effects could be more clearly understood from the focusing pattern plotted in Figure 6b–d at σ = 0°, 90° and 120°, respectively. The spatial characteristics of the zeroth diffraction order are hardly influenced when the phase distortion gets more deteriorated. Its intensity decreases as the unwanted higher diffraction orders become stronger. For practice, we can expect that the proposed off-axis meta-lens can suffer a spatial phase distortion with the largest deviation less than 48° corresponding to a 50% amplitude reduction where the higher diffraction orders are thought weakly enough and have no influence on the performance of the spectrometer. This is relatively a large tolerance and allows the device to be precisely implemented.

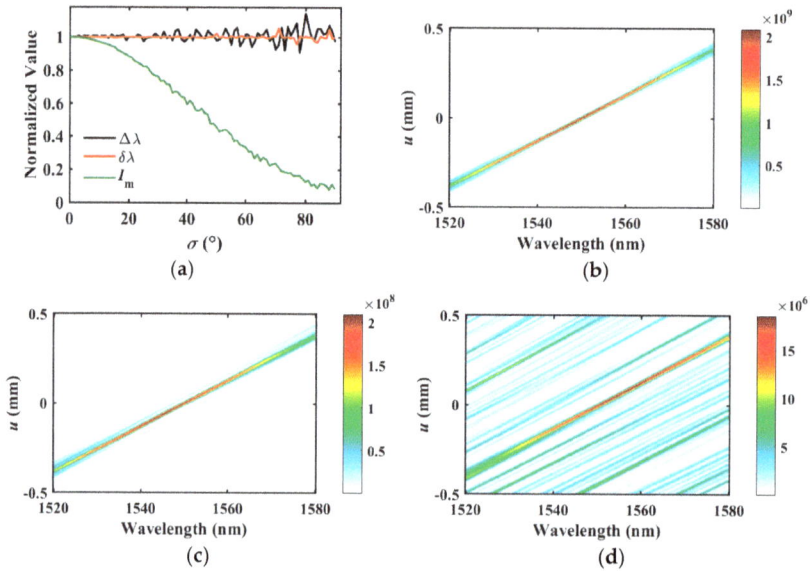

Figure 6. Influence of error analysis: (**a**) The effective spectral range $\Delta\lambda$ (black), the spectral resolution $\delta\lambda$ (red) and the maximum intensity I_m (green) of focal spot changing as a function of the random phase error factor σ. The values are normalized by those at $\sigma = 0$. Here the added spatial phase distortion value obeys a normal distribution $\mathcal{N}(0, \sigma^2)$ realized by the built-in random function in MATLAB; (**b–d**) The focus pattern of the proposed off-axis meta-lens at $\sigma = 0°$, $90°$ and $120°$, respectively.

4. Conclusions

In this paper, the relationship between the performance of the off-axis meta-lens-based spectrometer and the structural parameters of the system are studied theoretically. It is shown that the desired effective spectral range $\Delta\lambda$ and the spectral resolution $\delta\lambda$ can be realized by properly selecting the designing parameters of focal length f, aperture diameter D, off-axis angle α and orientation angle of output plane β. Based on these, we designed two device configurations able to work for (i) broadband spectroscopy in 800–1800 nm with the low spectral resolution of 2–5 nm and (ii) narrowband spectroscopy in 780–920 nm with the high spectral resolution of 0.15–0.6 nm. These results will be helpful to guiding the design of compact spectrometer with specific response characteristics for various applications.

Acknowledgments: The authors are grateful to the partial supports from NSFC of Zhejiang Province (LR15F050001& LZ17A040001), NSFC (61775195) and the National Key Research and Development Program of China (No. 2017YFA0205700).

Author Contributions: Yi Zhou performed the simulations, analyzed the data and wrote the paper; Rui Chen provided the main instructions of the simulation methods; Yungui Ma supervised this study and revised the paper.

Conflicts of Interest: The authors declare no conflict of interest.

References

1. Staude, I.; Schilling, J. Metamaterial-Inspired Silicon Nanophotonics. *Nat. Photonics* **2017**, *11*, 274–284. [CrossRef]
2. Liu, L.; Zhang, X.; Kenney, M.; Su, X.; Xu, N.; Ouyang, C.; Shi, Y.; Han, J.; Zhang, W.; Zhang, S. Broadband Metasurfaces with Simultaneous Control of Phase and Amplitude. *Adv. Mater.* **2014**, *26*, 5031–5036. [CrossRef] [PubMed]

3. Ni, X.; Emani, N.K.; Kildishev, A.V.; Boltasseva, A.; Shalaev, V.M. Broadband Light Bending with Plasmonic Nanoantennas. *Science* **2012**, *335*, 427. [CrossRef] [PubMed]
4. Pors, A.; Nielsen, M.G.; Eriksen, R.L.; Bozhevolnyi, S.I. Broadband Focusing Flat Mirrors Based on Plasmonic Gradient Metasurfaces. *Nano Lett.* **2013**, *13*, 829–834. [CrossRef] [PubMed]
5. Arbabi, A.; Horie, Y.; Bagheri, M.; Faraon, A. Dielectric Metasurfaces for Complete Control of Phase and Polarization with Subwavelength Spatial Resolution and High Transmission. *Nat. Nanotechnol.* **2015**, *10*, 937–943. [CrossRef] [PubMed]
6. Balthasar Mueller, J.P.; Rubin, N.A.; Devlin, R.C.; Groever, B.; Capasso, F. Metasurface Polarization Optics: Independent Phase Control of Arbitrary Orthogonal States of Polarization. *Phys. Rev. Lett.* **2017**, *118*, 113901. [CrossRef] [PubMed]
7. Khorasaninejad, M.; Crozier, K.B. Silicon Nanofin Grating as a Miniature Chirality-Distinguishing Beam-Splitter. *Nat. Commun.* **2014**, *5*, 5386. [CrossRef] [PubMed]
8. Arbabi, A.; Arbabi, E.; Kamali, S.M.; Horie, Y.; Han, S.; Faraon, A. Miniature Optical Planar Camera Based on a Wide-Angle Metasurface Doublet Corrected for Monochromatic Aberrations. *Nat. Commun.* **2016**, *7*, 13682. [CrossRef] [PubMed]
9. Wen, D.; Yue, F.; Ardron, M.; Chen, X. Multifunctional Metasurface Lens for Imaging and Fourier Transform. *Sci. Rep.* **2016**, *6*, 27628. [CrossRef] [PubMed]
10. Yang, H.; Li, G.; Su, X.; Cao, G.; Zhao, Z.; Chen, X.; Lu, W. Reflective Metalens with Sub-Diffraction-Limited and Multifunctional Focusing. *Sci. Rep.* **2017**, *7*, 12632. [CrossRef] [PubMed]
11. Khorasaninejad, M.; Zhu, A.Y.; Roques-Carmes, C.; Chen, W.T.; Oh, J.; Mishra, I.; Devlin, R.C.; Capasso, F. Polarization-Insensitive Metalenses at Visible Wavelengths. *Nano Lett.* **2016**, *16*, 7229–7234. [CrossRef] [PubMed]
12. Khorasaninejad, M.; Chen, W.T.; Zhu, A.Y.; Oh, J.; Devlin, R.C.; Rousso, D.; Capasso, F. Multispectral Chiral Imaging with a Metalens. *Nano Lett.* **2016**, *16*, 4595–6000. [CrossRef] [PubMed]
13. Chen, W.T.; Zhu, A.Y.; Khorasaninejad, M.; Shi, Z.; Sanjeev, V.; Capasso, F. Immersion Meta-Lenses at Visible Wavelengths for Nanoscale Imaging. *Nano Lett.* **2017**, *17*, 3188–3194. [CrossRef] [PubMed]
14. Groever, B.; Chen, W.T.; Capasso, F. Meta-Lens Doublet in the Visible Region. *Nano Lett.* **2017**, *17*, 4902–4907. [CrossRef] [PubMed]
15. Ni, X.; Kildishev, A.V.; Shalaev, V.M. Metasurface Holograms for Visible Light. *Nat. Commun.* **2013**, *4*, 2807. [CrossRef]
16. Chen, W.T.; Yang, K.Y.; Wang, C.M.; Huang, Y.W.; Sun, G.; Chiang, I.D.; Liao, C.Y.; Hsu, W.L.; Lin, H.T.; Sun, S.; et al. High-Efficiency Broadband Meta-Hologram with Polarization-Controlled Dual Images. *Nano Lett.* **2014**, *14*, 225–230. [CrossRef] [PubMed]
17. Wen, D.; Yue, F.; Li, G.; Zheng, G.; Chan, K.; Chen, S.; Chen, M.; Li, K.F.; Wong, P.W.H.; Cheah, K.W.; et al. Helicity Multiplexed Broadband Metasurface Holograms. *Nat. Commun.* **2015**, *6*, 8241. [CrossRef] [PubMed]
18. Zheng, G.; Mühlenbernd, H.; Kenney, M.; Li, G.; Zentgraf, T.; Zhang, S. Metasurface Holograms Reaching 80% Efficiency. *Nat. Nanotechnol.* **2015**, *10*, 308–312. [CrossRef] [PubMed]
19. Zhang, C.; Yue, F.; Wen, D.; Chen, M.; Zhang, Z.; Wang, W.; Chen, X. Multichannel Metasurface for Simultaneous Control of Holograms and Twisted Light Beams. *ACS Photonics* **2017**, *4*, 1906–1912. [CrossRef]
20. Khorasaninejad, M.; Chen, W.T.; Oh, J.; Capasso, F. Super-Dispersive Off-Axis Meta-Lenses for Compact High Resolution Spectroscopy. *Nano Lett.* **2016**, *16*, 3732–3737. [CrossRef] [PubMed]
21. Zhu, A.Y.; Chen, W.T.; Khorasaninejad, M.; Oh, J.; Zaidi, A.; Mishra, I.; Devlin, R.C.; Capasso, F. Ultra-Compact Visible Chiral Spectrometer with Meta-Lenses. *APL Photonics* **2017**, *2*, 036103. [CrossRef]
22. Minovich, A.E.; Miroshnichenko, A.E.; Bykov, A.Y.; Murzina, T.V.; Neshev, D.N.; Kivshar, Y.S. Functional and Nonlinear Optical Metasurfaces. *Laser Photonics Rev.* **2015**, *9*, 195–213. [CrossRef]
23. Walter, F.; Li, G.; Meier, C.; Zhang, S.; Zentgraf, T. Ultrathin Nonlinear Metasurface for Optical Image Encoding. *Nano Lett.* **2017**, *17*, 3171–3175. [CrossRef] [PubMed]
24. Zhang, L.; Mei, Z.L.; Zhang, M.R.; Yang, F.; Cui, T.J. An Ultrathin Directional Carpet Cloak Based on Generalized Snell's Law. *Appl. Phys. Lett.* **2013**, *103*, 151115. [CrossRef]
25. Ni, X.; Wong, Z.J.; Mrejen, M.; Wang, Y.; Zhang, X. An Ultrathin Invisibility Skin Cloak for Visible Light. *Science* **2015**, *349*, 1310–1314. [CrossRef] [PubMed]

26. Yang, Y.; Jing, L.; Zheng, B.; Hao, R.; Yin, W.; Li, E.; Soukoulis, C.M.; Chen, H. Full-Polarization 3D Metasurface Cloak with Preserved Amplitude and Phase. *Adv. Mater.* **2016**, *28*, 6866–6871. [CrossRef] [PubMed]

27. Born, M.; Wolf, E. *Principles of Optics: Electromagnetic Theory of Propagation, Interference and Diffraction of Light*, 7th ed.; Cambridge University Press: Cambridge, UK, 1999, ISBN 978-0-521-64222-4.

28. Bao, J.; Bawendi, M.G. A Colloidal Quantum Dot Spectrometer. *Nature* **2015**, *523*, 67–70. [CrossRef] [PubMed]

29. Henstridge, M.; Zhou, J.; Guo, L.J.; Merlin, R. Wavelength Scale Terahertz Spectrometer Based on Extraordinary Transmission. *Appl. Phys. Lett.* **2017**, *111*, 063503. [CrossRef]

30. Yuan, L.; He, Z.; Lv, G.; Wang, Y.; Li, C.; Xie, J.; Wang, J. Optical Design, Laboratory Test, and Calibration of Airborne Long Wave Infrared Imaging Spectrometer. *Opt. Express* **2017**, *25*, 22440–22454. [CrossRef] [PubMed]

31. Cook, T. Anamorphic Integral Field Spectrometer for Diffuse Ultraviolet Astronomy. *Appl. Opt.* **2013**, *52*, 8765–8770. [CrossRef] [PubMed]

32. Cataldo, G.; Hsieh, W.-T.; Huang, W.-C.; Moseley, S.H.; Stevenson, T.R.; Wollack, E.J. Micro-Spec: An Ultracompact, High-Sensitivity Spectrometer for Far-Infrared and Submillimeter Astronomy. *Appl. Opt.* **2014**, *53*, 1094–1102. [CrossRef] [PubMed]

33. Zavvari, A.; Islam, M.T.; Anwar, R.; Abidin, Z.Z.; Asillam, M.F.; Monstein, C. Analysis of Radio Astronomy Bands Using CALLISTO Spectrometer at Malaysia-UKM Station. *Exp. Astron.* **2016**, *41*, 185–195. [CrossRef]

34. Udeigwe, T.K.; Young, J.; Kandakji, T.; Weindorf, D.C.; Mahmoud, M.A.; Stietiya, M.H. Elemental Quantification, Chemistry, and Source Apportionment in Golf Course Facilities in a Semi-Arid Urban Landscape Using a Portable X-Ray Fluorescence Spectrometer. *Solid Earth* **2015**, *6*, 415–424. [CrossRef]

35. Buzan, E.M.; Beale, C.A.; Boone, C.D.; Bernath, P.F. Global Stratospheric Measurements of the Isotopologues of Methane from the Atmospheric Chemistry Experiment Fourier Transform Spectrometer. *Atmos. Meas. Tech.* **2016**, *9*, 1095–1111. [CrossRef]

36. Mantouvalou, I.; Lachmann, T.; Singh, S.P.; Vogel-Mikuš, K.; Kanngießer, B. Advanced Absorption Correction for 3D Elemental Images Applied to the Analysis of Pearl Millet Seeds Obtained with a Laboratory Confocal Micro X-ray Fluorescence Spectrometer. *Anal. Chem.* **2017**, *89*, 5453–5460. [CrossRef] [PubMed]

37. Shaltout, A.; Liu, J.; Kildishev, A.; Shalaev, V. Photonic Spin Hall Effect in Gap Plasmon Metasurfaces for On-Chip Chiroptical Spectroscopy. *Optica* **2015**, *2*, 860–863. [CrossRef]

38. Maguid, E.; Yulevich, I.; Veksler, D.; Kleiner, V.; Brongersma, M.L.; Hasman, E. Photonic Spin-Controlled Multifunctional Shared-Aperture Antenna Array. *Science* **2016**, *352*, 1202–1206. [CrossRef] [PubMed]

39. Ding, F.; Pors, A.; Chen, Y.; Zenin, V.A.; Bozhevolnyi, S.I. Beam-Size-Invariant Spectropolarimeters Using Gap-Plasmon Metasurfaces. *ACS Photonics* **2017**, *4*, 943–949. [CrossRef]

40. Zhou, Y.; Chen, R.; Ma, Y. Design of Optical Wavelength Demultiplexer Based on Off-Axis Meta-Lens. *Opt. Lett.* **2017**, *42*, 4716–4719. [CrossRef] [PubMed]

41. Goodman, J.W. *Introduction to Fourier Optics*, 3rd ed.; Roberts & Company Publishers: Englewood, IL, USA, 2005, ISBN 978-0-9747077-2-3.

42. Noll, R.J. Zernike Polynomials and Atmospheric Turbulence. *J. Opt. Soc. Am. A* **1976**, *66*, 201–211. [CrossRef]

43. Welford, W.T. *Aberrations of Optical Systems*, 1st ed.; IOP Publishing: Bristol, UK, 1986, ISBN 978-0-85274-564-9.

44. Hamamatsu Mini-Spectrometers Product Page. Available online: http://www.hamamatsu.com/eu/en/product/category/5001/4016/index.html (accessed on 29 November 2017).

applied
sciences

MDPI

Article

Waveguide Coupling via Magnetic Gratings with Effective Strips

Kevin M. Roccapriore [1], David P. Lyvers [1,2], Dean P. Brown [2,3], Ekaterina Poutrina [2,3], Augustine M. Urbas [3], Thomas A. Germer [4] and Vladimir P. Drachev [1,5,*]

[1] Department of Physics and Advance Materials Manufacturing Processing Institute, University of North Texas, Denton, TX 76203, USA; KevinRoccapriore@my.unt.edu (K.M.R.); walkingcub@gmail.com (D.P.L.)

[2] UES, Inc., 4401 Dayton-Xenia Rd, Dayton, OH 45432, USA; dbrown@ues.com (D.P.B.); ekaterina.poutrina.ctr.ru@us.af.mil (E.P.)

[3] Air Force Research Lab, Materials and Manufacturing Directorate, 3005 Hobson Way, Wright Patterson AFB, Dayton, OH 45433, USA; augustine.urbas@wpafb.af.mil

[4] Sensor Science Division, National Institute of Standards and Technology, 100 Bureau Drive, Gaithersburg, MD 20899, USA; thomas.germer@nist.gov

[5] Center of Photonics & Quantum Materials, Skolkovo Institute of Science and Technology, 121205 Moscow, Russia

* Correspondence: vladimir.drachev@unt.edu; Tel.: +1-940-565-4580

Received: 20 February 2018; Accepted: 10 April 2018; Published: 14 April 2018

Abstract: Gratings with complex multilayer strips are studied under inclined incident light. Great interest in these gratings is due to applications as input/output tools for waveguides and as subwavelength metafilms. The structured strips introduce anisotropy in the effective parameters, providing additional flexibility in polarization and angular dependences of optical responses. Their characterization is challenging in the intermediate regime between subwavelength and diffractive modes. The transition between modes occurs at the Wood's anomaly wavelength, which is different at different angle of incidence. The usual characterization with an effective film using permittivity ε and permeability μ has limited effectiveness at normal incidence but does not apply at inclined illumination, due to the effect of periodicity. The optical properties are better characterized with effective medium strips instead of an effective medium layer to account for the multilayer strips and the underlying periodic nature of the grating. This approach is convenient for describing such intermediate gratings for two types of applications: both metafilms and the coupling of incident waves to waveguide modes or diffraction orders. The parameters of the effective strips are retrieved by matching the spectral-angular map at different incident angles.

Keywords: metamaterials; homogenization; magnetic grating; waveguide coupling; metasurfaces

1. Introduction

There are two different applications of gratings in general. First is a diffraction tool with a period larger than the wavelength, and second is as an engineered film with controlled material parameters. The second type of application requires substantially subwavelength gratings, so that in some ranges of wavelengths and angles they can be described by the effective parameters of a uniform film. For those ranges of wavelengths and angles there are no detectable non-zero diffraction orders. One-dimensional gratings consisting of stacked metal-dielectric strips are investigated for their ability to provide magnetic as well as electric resonances [1–4]. Such resonances are located in the visible spectrum due to the size of the unit cell—less than 0.5 μm. Effective magnetic permeability appears due to circular currents in the stacked structure. It follows from Maxwell's equations that the transmission and reflection coefficients for the effective film depend on both the product and the ratio of permittivity and permeability, i.e., on the refractive index and the impedance [5,6]. As for any nonlocal effect, the magnetic response increases with the size-to-wavelength ratio. That is why, in the case of metamaterial applications, we are often at the borderline of applicability of the effective parameter approach [7,8]. Often, the grating can be treated as an effective film at normal incidence but it cannot be at inclined incidence. The boundary angle at a specific wavelength is defined by appearance of the first diffraction order, which is the Wood's anomaly [9–11].

Here we characterize such a grating consisting of a stacked vertical substructure by an effective permittivity and permeability of the strips instead of by a continuous effective layer. The approach captures both the effect of the substructure resulting in an artificial permeability, and the diffraction of a grating, by employing a set of effective parameters for the geometrically defined strips. It should be clear that, as is the case with any effective medium approximation, the model works best in the long wavelength limit compared to the strip dimensions. Note, that the effective layer approach can be applicable only in the spectral range of the zero-order diffraction. Any high-order diffraction would result in a wrong retrieval. The experiments presented here show first diffraction order at 450 nm making effective layer approximation not applicable. It is important to note that a grating with the structured strips can be of great interest for the first type of application, especially as an input/output tool for waveguides. Indeed, the structured strips introduce anisotropy in the effective parameters, which makes it possible to realize different polarization and angular dependences, and this behavior is captured by the proposed model.

For similar grating structures, the permeability and permittivity retrieval process has been previously demonstrated for normal-incidence illumination using rigorous coupled wave (RCW) methods [5,6,12] and using complex transmission and reflection coefficients [3,4]. The retrieval for the effective films, however, cannot be verified conclusively for inclined illumination. This is in large part due to the difficulty arising from the effect of diffraction [8,9]. While diffraction occurs at normal incidence, its higher order effects in the range from 400 nm to 1000 nm are typically masked in the effective layer scheme. Under inclined illumination, the Wood's anomaly is red-shifted into the visible, higher diffraction orders are no longer hidden, and the effective layer method breaks down.

2. Materials and Methods

In this paper, we study a stacked metal-dielectric grating, shown schematically in Figure 1a, at inclined illumination and develop a method to retrieve an effective anisotropic permittivity and permeability of the strips. The grating is fabricated on a transparent substrate and illuminated with white light at various angles of incidence, while light diffracted from the grating is partially coupled to the waveguiding mode and collected as a function of angle from the edge of the substrate, as shown in Figure 1c. The collected light is then delivered to the spectrometer by a fiber bundle. Additionally, normal incidence transmission and reflection measurements are performed in order to provide some information for spectral positions of resonances. The setup described in Figure 1c will produce a wavelength-angle intensity map. This map will be unique for each angle of incidence and each polarization. The simulation efforts will produce the same intensity maps and transmission spectra to match the experimental spectra by adjusting the unknown strip parameters. Thus, using an iterative procedure where the strip parameters are slightly adjusted individually until the best goodness of fit is obtained, we can determine the proper set of parameters. The fitting is done manually, such that the simulated data appears to visually match the experimental data. This matching process is further detailed in the Results and Discussion section. Here we choose modified rigorous coupled wave (RCW) analysis which is utilized as a package within the Modeled Integrated Scattering Tool (MIST [13]). MIST is a front-end graphical user interface to the SCATMECH C++ library of scattering codes, based on rigorous coupled wave theory [14–16], modified to account for anisotropic permittivity and permeability (see Appendix A). With this tool, we are able to model diffraction effects of any order. The matching process must be done for each incident angle and each polarization, including that of normal incidence transmission, and the same unique set of permittivity and permeability functions is required for successful matching. The simulation is quite sensitive to the spectral position, amplitude, and line-shape of the strips' permittivity and permeability functions. The beauty of this approach also lies in the fact that we need not be concerned with the artifacts in the retrieval caused by the grating periodicity.

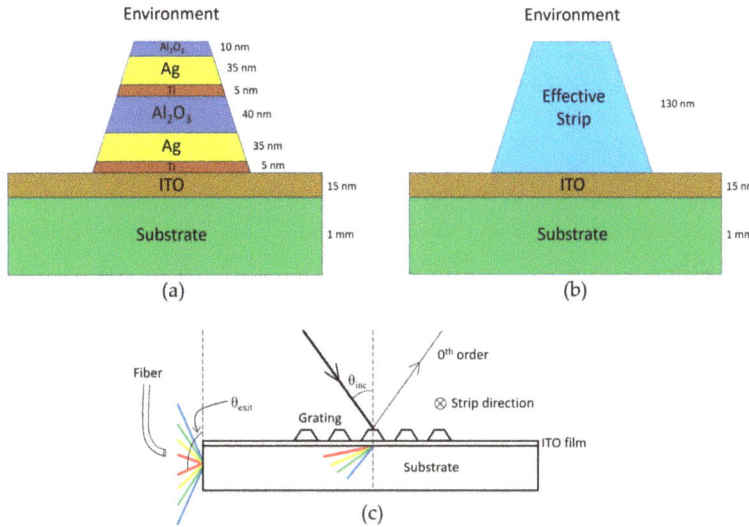

Figure 1. Sample geometry and experimental setup. Panel (**a**) shows the vertical substructure of the strip; (**b**) portrays the replacement of the real structure with an effective strip; (**c**) depicts the inclined illumination setup, in addition to usual far-field transmission spectroscopy, necessary for retrieving the optical parameters.

Any isotropic medium can be optically characterized by either pair of parameters, index of refraction n and impedance Z, or by permittivity ε and permeability μ. All four quantities are causal, complex, and depend on frequency ω. They are related by

$$n = \sqrt{\varepsilon\mu} \text{ and } Z = \sqrt{\frac{\mu}{\varepsilon}} \tag{1}$$

In accounting for a permeability different from unity, it is clear the permittivity and permeability are treated as complex quantities obeying the Kramers–Kronig relations and n and Z are independent functions.

One approach to determine the values for ε and μ of a material, sometimes termed the "effective layer method," is to use RCW in which the incident light is normal upon the structure. This method however considers the grating as a continuous film and hence, the properties of an effective continuous layer are retrieved. Here, we seek a more complete retrieval which is not restricted to normal incidence, and thus accounts for anisotropy. In this more general situation of incline incidence, a redshift of the Wood's anomaly into the visible range occurs, which is precisely the wavelength region of our interest and the effective layer retrieval has limited practical use here.

Due to the periodic nature and size of the structure relative to the incident wavelength, an electromagnetic plane wave will undergo diffraction and will transfer some of its power into higher orders. Diffraction effects can be described by the well-known grating equation

$$n \sin(\theta_m) = n_{inc}\sin(\theta_{inc}) - \frac{m\lambda}{p}, \tag{2}$$

where the subscript-free n refers to the transmitted region's refractive index, n_{inc} is the refractive index in the incident medium, p is the period of the grating, λ is the wavelength in vacuum, and m is the diffraction order. For the grating under study, our substrate also serves as a waveguide. We wish to simulate the diffraction that occurs due to a periodic grating structure, and the MIST suits our needs for this. MIST includes RCW analysis to simulate the interaction of arbitrarily polarized light with a grating structure of interest. More detail regarding MIST will be given in the subsequent simulation subsection.

The strips of the diffraction grating consist of a total of six layers, essentially forming a metal–insulator–metal (MIM) configuration, which includes adhesion and oxide protection layers. As depicted in Figure 1, each layer of silver lays atop a thin titanium layer, with the metal layers separated by a spacer dielectric, and finally a top protective oxide layer. The die size is 500 μm × 500 μm, with a period of 305 nm and a total height of 130 nm. The bottom and top widths are 155 nm and 80 nm, respectively, which causes an asymmetry to the shape of the two metal strips such that the bottom silver layer is wider than the top silver layer. The metal strips themselves have a thickness of 35 nm, with a 40 nm layer of alumina separating them. Silver is the selected metal due to its low losses at optical frequencies, while alumina has been chosen as the spacer dielectric for its high dielectric constant. It has been shown that the higher dielectric constant spacer is more suitable for magnetic grating metamaterials because it provides better field confinement [17]. Samples have been fabricated on a 15-nm indium-tin-oxide- (ITO) coated fused silica substrate using conventional electron beam lithography (EBL) techniques. The ITO layer is used primarily to provide conduction during EBL. Note that a trapezoidal shape of the cross-section is due to the applied fabrication protocol [6]. After development of the exposed resist, titanium, silver, and alumina layers are deposited by electron beam evaporation. A 3-nm layer of titanium is evaporated before each silver layer to provide good adhesion, making the samples more robust, but lowering the quality of the plasmonic resonances. It is worth mentioning that the roughness and grating quality have been found to be significantly affected by deposition rate [18]. This in turn can affect the optical characteristics of the sample. Specifically, lower deposition rates for gratings and other finer features tend to yield smoother and better-quality nanostructures, as opposed to higher deposition rates giving better quality continuous films. As such, a low deposition rate of 0.1 nm/s has been used. A final liftoff process in acetone is performed revealing the intended grating structure. Figure 2 shows a scanning electron microscopy (SEM) micrograph of the top view of the sample.

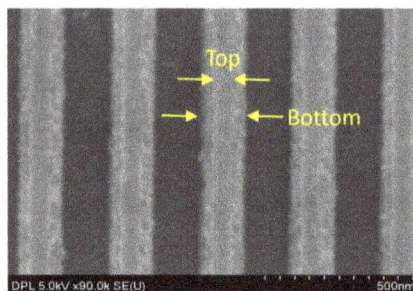

Figure 2. Scanning electron microscopy (SEM) micrograph of meta-grating sample. Note the trapezoidal shape.

Upon successful fabrication, the sample is used in conjunction with the optical scheme described in Figure 1c. Under normal incidence illumination, diffraction does occur for shorter wavelengths; however, the nonzero orders are trapped in the glass substrate by total internal reflection, and as a result, the detector only detects zero-order diffraction. Supercontinuum white light pulses converted from 800-nm pump (Figure 3) of either transverse electric (TE) or transverse magnetic (TM) polarization illuminate the sample at differing angles of incidence, relative to the sample surface normal. The spectra of the transmission through the substrate at incline incidence are used to normalize the spectral response of the samples. Figure 3 gives an example of such spectra at 30°. The laser source stability is

10% and an example of its output at a particular angle through the substrate as well as its stability with time and with different polarizations is shown in Figure 3. Upon striking the grating, light diffracts at many angles. These diffracted rays are waveguided by total internal reflection through the substrate only in the negative direction. The intensities are collected via a scanned fiber of core diameter 500 μm located a distance of 0.5 mm from the substrate edge, and is subsequently delivered to an imaging spectrometer, whose spectral resolution is 1.5 nm. Note the angular resolution of measurement is 0.5° The output intensity will then be a function of both the angle and wavelength in a spectral-angular intensity distribution (Figure 1c). We ignore rays that propagate in the positive direction which may eventually return to the detector by means of many internal reflections. The reason for this is that, with the angles of incidence used, these rays will *not* undergo total internal reflection within the substrate, and due to significant loss from many of these repeated reflections, they contribute several orders of magnitude less signal. The process is repeated for incident angles of 30°, 40°, 50°, and 60°, and for two linear polarizations (TE or TM) for a total of eight spectral-angular intensity maps. What follows is a matching process utilizing simulation methods for both normal incidence transmission (zero-order diffraction) as well as the spectral-angular map (nonzero-order diffraction).

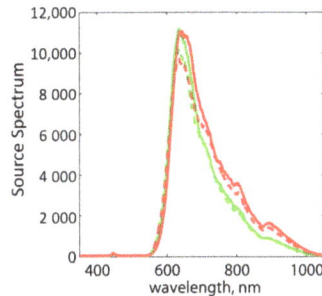

Figure 3. Substrate transmission at 30° incline, showing the stability of the source output spectra. These spectra are collected for each angle of incidence and used later for normalizing procedures. Green and red lines represent TM and TE polarizations, respectively. The dashed variants show the stability of each after one hour.

Note, that in past retrieval schemes, the grating is approximated as an effective layer and far field transmission and reflection simulations are matched to experimentally observed data (for example [1–3]). The type of simulation, that is based on RCW methods, allows retrieval of the complex transmission and reflection coefficients, which can then in turn be used to calculate the permittivity and permeability, albeit as a continuous effective layer. In reality, since a periodic structure is used, diffraction occurs at normal incidence for $\lambda \leq np$ where n is the substrate refractive index. In our case, this approximately translates to first order diffraction occurring at normal incidence for wavelengths less than 450 nm. Thus the effective layer approximation is valid only for the spectral range where the measured transmittance/reflection only contains zero-order information and the higher (nonzero) orders are not allowed.

In this paper, retrieval of the individual strips' effective optical properties is accomplished by matching simulation spectral-angular data from the designed waveguiding structure, as well as the normal incidence transmission, to the corresponding set of experimental data. For this we use the RCW model included in the MIST. To describe the interacting system, MIST requires the grating geometry, substrate and superstrate media, incident light wavelength, polarization, and angle of incidence, as well as the optical properties of the strips. Despite the strips consisting of several metal and dielectric layers, we model the strip as a single unit—the effective strip. All the roughness and crystal quality of the materials are included in the effective parameters. The physical dimensions of the strip are those of the real sample, while the effective parameters of the strips are what we seek, and are also the only set of unknown variables. A unique ability of MIST is that it allows the calculation of any arbitrary diffraction order as well as anisotropic magnetic behavior. In this way, we can provide any tabulated dispersion for a range of frequencies for all permittivity and permeability functions. We briefly mention that, for inclined illumination and the wavelengths we use, it is only necessary to analyze a single diffraction order. Namely $m = -1$, the first order, contributes to a meaningful relative intensity at the output. Minus is due to geometrical convention. The other orders either do not exist, or their efficiencies are negligible—as is the case for orders higher than the first order.

Any arbitrary complex function for both $\varepsilon(\lambda)$ and $\mu(\lambda)$ can be supplied to MIST. Typically, in anisotropic media the electric (magnetic) susceptibility and therefore permittivity (permeability) functions exist in the form of a second rank tensor. Each function will uniquely have three nonzero effective medium components, namely:

$$\varepsilon = \begin{pmatrix} \varepsilon_x & 0 & 0 \\ 0 & \varepsilon_y & 0 \\ 0 & 0 & \varepsilon_z \end{pmatrix} \quad and \quad \mu = \begin{pmatrix} \mu_x & 0 & 0 \\ 0 & \mu_y & 0 \\ 0 & 0 & \mu_z \end{pmatrix}, \tag{3}$$

where each component is a complex and frequency dependent quantity. The off-diagonal components are zero due to the geometry of the grating. Figure 4 shows the coordinate system being used.

If only normal incidence transmission is utilized, it is not guaranteed that all the components of the effective permittivity and permeability of the strips are involved. We find that there are several functions that will provide a suitable match to the transmission data. Additionally, the z-components of the permittivity and permeability are concealed at normal incidence, and we only begin to noticeably detect their effect at larger angles of incidence. Only when the normal incidence matching is used in conjunction with the incline illumination results can we obtain the correct set of effective optical properties of the system.

We next discuss the physical grounds for the accepted fitting formulas on the six unknown components. Different polarizations and incident angles are used to isolate and better capture specific components of the effective permittivity and permeability. Moreover, we find there is a significant polarization dependency on the output, due to the strong resonance occurring with the TM polarization and lack of resonance with TE polarization. Figure 4 illustrates that when the E-field lies along the strip axis, this TE-polarized wave is unable to produce any resonant effects, and thus the strip will behave as a diluted metal. In this case the observable permeability is unity, hence, $\mu_x = \mu_z = 1$. We note that by keeping the x- and z-components of the effective permeability, nonmagnetic response is an enforced condition. Meanwhile the observable component when using TE polarization will be strictly that of ε_y, though as we shall see it differs moderately from the standard EMT of a dilute metal. Hence by using TE polarized light, we can isolate the y-component of the permittivity.

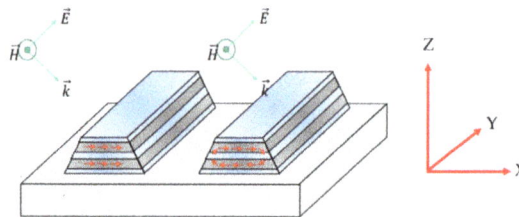

Figure 4. Representation of the actual trapezoidal shape of the grating due to fabrication limitations. A TM-polarized beam is incident at an incline; depending on frequency, either a symmetric (**Left**) or antisymmetric (**Right**) mode may be excited. Note that the strips are still considered to be infinite in the y-direction.

On the other hand, when the incident wave is TM polarized, both symmetric and antisymmetric resonant current modes may be excited. Here we see no effect from either μ_x or μ_z due to the fixed direction of the magnetic field along the strip. However, a magnetic dipole response manifests with μ_y via the oscillations of the light wave's magnetic field. In fact, it is this magnetic resonance which is precisely the desired effect that we hope to observe. The electric permittivity takes form with ε_x and ε_z. It should be clear that with normal incidence measurements, the total anisotropy cannot be recovered. Let us now summarize the results provided by polarization:

$$\varepsilon = \begin{bmatrix} 0 & 0 & 0 \\ 0 & \varepsilon_y & 0 \\ 0 & 0 & 0 \end{bmatrix} + \begin{bmatrix} \varepsilon_x & 0 & 0 \\ 0 & 0 & 0 \\ 0 & 0 & \varepsilon_z \end{bmatrix},$$
$$\mu = \begin{bmatrix} 1 & 0 & 0 \\ 0 & 0 & 0 \\ 0 & 0 & 1 \end{bmatrix} + \begin{bmatrix} 0 & 0 & 0 \\ 0 & \mu_y & 0 \\ 0 & 0 & 0 \end{bmatrix},$$
(4)

where the left matrices are sensed with TE polarization and the right matrices are sensed with TM polarization.

It is only ε_x and μ_y which display resonant behavior, while ε_y shows dilute metal characteristics, and μ_x and μ_z are simply unity. The z-component of the permittivity is not expected to be resonant in the visible spectrum, though its form via simulation will turn out to be that of a somewhat modified Drude function.

There are, then, a total of four unknown parameters that must be modeled. For the TM resonant modes, both the permittivity and permeability functions have asymmetry of the resonance (see, for example [5,18]). Because of this asymmetry, we choose a transversal optical longitudinal optical (TOLO) oscillator function [19] to describe these modes, rather than a classical Lorentzian. These take the form

$$\varepsilon_x = A_{\varepsilon_x} \frac{\omega_{LO}^2 - \omega^2 - i\Gamma_{LO}\omega}{\omega_{TO}^2 - \omega^2 - i\Gamma_{TO}\omega} \tag{5}$$

and

$$\mu_y = A_{\mu_y} \frac{\omega_{LO}^2 - \omega^2 - i\Gamma_{LO}\omega}{\omega_{TO}^2 - \omega^2 - i\Gamma_{TO}\omega} \tag{6}$$

where, after matching, we find $A_{\varepsilon_x} = 0.11$, $\omega_{LO} = 4.5$ eV, $\omega_{TO} = 2.17$ eV, $\Gamma_{LO} = 1.5$ eV, and $\Gamma_{TO} = 0.3$ eV, while for μ_y, $A_{\mu_y} = 0.13$, $\omega_{LO} = 4.0$ eV, $\omega_{TO} = 1.77$ eV, $\Gamma_{LO} = 2.0$ eV, and $\Gamma_{TO} = 0.4$ eV. Here, $\omega = hc/\lambda$ is the photon energy. For these dielectric functions to remain physical with $\text{Im}\{\varepsilon_x\} \geq 0$ and $\text{Im}\{\mu_y\} \geq 0$, the constraint $\Gamma_{LO} - \Gamma_{TO} > 0$ must be satisfied.

Meanwhile, to model "diluted metal" for both ε_y and ε_z, we use a Drude function of the form

$$\varepsilon_{y,z} = \varepsilon_{\infty y,z} - \frac{A_{y,z}\omega_p^2}{\omega^2 + i\Gamma_{y,z}\omega} \tag{7}$$

where $\omega_p = 9$ eV and for ε_y we have $A_y = 0.07$, $\Gamma_y = 4$ eV, and $\varepsilon_{\infty y} = 3.5$, while for ε_z we have $A_z = 0.123$, $\Gamma_z = 1$ eV, and $\varepsilon_{\infty z} = 9$. It is interesting to note that without the large offset parameter, $\varepsilon_{\infty z}$, for ε_z the Wood's anomaly peak will be hidden in transmission spectra.

The details regarding the exact spectral position, amplitude, and sharpness of each resonance and the parameters for the dielectric functions are initially unknown. Therefore, an iterative technique must be performed until an acceptable match has been made for both normal incidence transmission as well as for each angle of incidence and polarization of the spectral angular map. As previously mentioned, TE polarization is employed to uniquely determine ε_y. As an example, for this case there are three parameters from the Drude relationship above that must be found. Values for each parameter are initially chosen with some physical justification, and subsequently the simulation is completed for all angles of incidence. The relative intensities of the spots on the spectral-angular map are analyzed, and one at a time the parameters are changed gradually to provide a closer match in intensity. The same parameters must also satisfy the normal incidence transmission for this polarization. In this way, we find the values that describe the y-component of the permittivity.

On the contrary, the TM polarization case is considerably more complex due to the fact that not one, but three components of permittivity and permeability affect the outcome, namely ε_x, ε_z, and μ_y. Additionally, the resonant functions each contain five fitting parameters resulting in several more degrees of freedom. However, there exist two resonances in the transmission spectra (Figure 5). The shorter wavelength resonance is associated with electric permittivity function and the longer wavelength resonance with magnetic permeability function. The following section provides details of the results of this matching procedure.

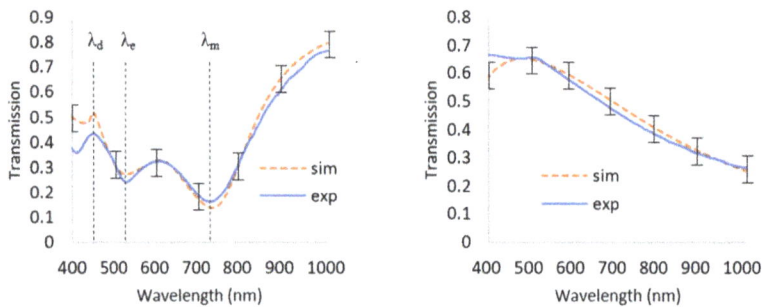

Figure 5. Normal incidence transmission for TM (**Left**) and TE (**Right**) polarizations. Note that only ε_y is responsible for TE spectra, while TM spectra depends on ε_x, ε_z, and μ_y. With TM polarization, λ_d, λ_e, and λ_m correspond to the Wood's anomaly (diffraction threshold), electric resonance, and magnetic resonance, respectively. Data is matched by providing incremental adjustments to the parameters of each dielectric function. Note that the set of parameters providing agreement here must also provide satisfactory matching for the spectral-angular data. Error bars in simulated spectra reflect $\pm 5\%$ uncertainty.

3. Results and Discussion

The designed grating is considered optically magnetic due to the ability of TM-polarized visible light to excite asymmetric circulating currents in the metal layers, thereby giving rise to a magnetic moment directly along the strips of the structure. This is fundamentally a result of the oscillating magnetic field along the strip axis and Faraday's law. Both a symmetric current mode and an asymmetric mode may be excited, representing an electric and magnetic resonance, respectively. The spectral position of the magnetic resonance has been previously demonstrated to be a result of the effective width of the grating (or fill factor) [20], though the normal incidence transmission local minima provide a baseline of sorts to determine the spectral locations for both ε_x and μ_y. As seen in Figure 5 the TM polarized transmission displays two local minima; the first, $\lambda_e = 530$ nm and the second, $\lambda_m = 725$ nm. These represent the electric and magnetic resonances, respectively, and specifically refer to ε_x and μ_y. Therefore, we have in mind a starting point for the spectral location of the resonances of the permittivities and permeability. The sharp peak at $\lambda_d = 450$ nm is due to the Wood anomaly and is not observed in simulation until ε_z is properly determined. As an example, if ε_z is given a constant nonabsorbing value of $\varepsilon_{\infty,z} = 1$, the sharp peak at λ_d is washed out; in this case a static offset to the real part of approximately $\varepsilon_{\infty,z} = 8$ is required for the diffraction threshold peak to appear.

The simulation results for the spectral-angular map are performed with the experimental values of the period of the grating, the location of the incident beam relative to the waveguide edge, as well as the substrate thickness and refractive index (see Figure 6). In Figure 6 we also explain why the spectral angular maps to be seen in Figures 7 and 8 are not continuous. Indeed, depending on the incident beam positions relative to the substrate edge, there are three possible scenarios. The beam at a particular wavelength may either hit the corner and split between upward and downward as shown in Figure 6a, or propagate in one of two directions, upward or downward. If the beam goes upward it makes a gap in the down side as indicated by arrows on Figure 6b. We emphasize that these aforementioned parameters taken from experiments control only the location and the size of the "spots" seen in the spectral-angular maps in Figures 7 and 8. More importantly, the permittivities and permeabilities are mainly responsible for the intensities of these spots.

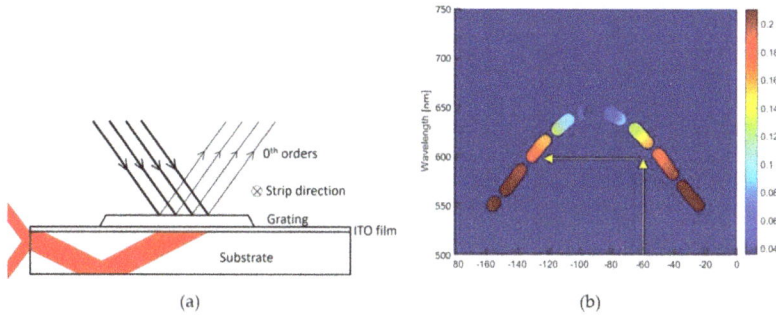

(a)   (b)

Figure 6. The geometry of the sample affects the output; this is shown for parallel rays interacting with the grating (**a**). Note, that depending on the incident beam positions relative to the substrate edge there are three possible scenarios, these rays may either hit the corner and split between upward and downward as shown in (**a**), or propagate in one of two directions, upward or downward. If the beam goes upward it makes a gap in the down side (**b**). Map (**b**) is an angular extension of Figure 8b SIM TM 40°, as an example; the gap at −60° is pointed by arrow.

To retrieve the effective optical properties, we begin with an initial guess for each parameter in the TOLO and Drude wavelength-dependent functions. This guess is influenced by the experimental transmission data and gives a starting point for the resonant spectral positions as well as the non-resonant spectra modeled as a diluted metal. These functions provide a value for permittivity and permeability at each wavelength, and are then fed to the MIST GUI which operates on each value using the modified RCW code. By varying the conditions, it will output either a transmission spectrum at normal incidence or a spectral-angular map—both must be performed. One by one, each parameter of the TOLO and Drude functions are iteratively adjusted until both the transmission and spectral-angular maps at all angles both match satisfactorily based on eye evaluation. This process is guided by the most sensitive parameters for the resonant spectral position, amplitude terms, followed by the spectral width term, and lastly the asymmetrical parameters. For example, to obtain μ_y, ε_x, and ε_z, we use the TM experimental data. Realizing that μ_y is responsible for the magnetic resonance and ε_x mainly responsible for the electric resonance, both amplitude and spectral position terms are initially chosen such that they best match the transmission data for the respective minima. The spectral positions for the resonant functions are initially chosen to be the same as those of the transmission minima. Note, these positions may not exactly coincide after finalizing the matching. Since the two resonances are not spectrally separated by a significant amount, increasing the amplitude of, for example, μ_y can have an impact on the simulated transmission's local *electric* minima, and vice versa.

Matching is assessed for the spectral-angular maps by comparing each spot's relative intensity. Upon doing so, we have best matched the simulated spectra to the experimental data. Thus, we have found each component previously discussed, namely, ε_x, ε_y, ε_z, and μ_y. Meanwhile μ_x and μ_z are set equal to unity as an enforced condition. Again, note that physical arguments are used to evaluate suitable functions, such as ε_y exhibiting a behavior similar to that of a dilute metal—a result of the non-resonant TE mode. The result of the normal incidence matching is shown in Figure 5, while all spectral-angular matching results for varied angles and polarizations are shown in Figures 7a–d and 8a–d. The results in Figures 7 and 8 all account for uncertainty in the spectral-angular position by allowing each data point to have a specific radius, such that it reflects the experimental data uncertainty. Furthermore, based on Figures 9 and 10, Figure 11 shows the functions chosen to satisfy matching of the experimental data. It is emphasized that the matching in Figures 5 and 7 and Figure 8 are not independent of one another, but rather the same set of permittivity and permeability must provide agreement for both.

To obtain a successful match, both normal incidence transmission and waveguided inclined illumination simulation data should agree with the experimental results. It is extremely challenging to simultaneously have both agree with a high degree of accuracy, especially using singly resonant Lorentzian-type functions with a limited number of fitting parameters. An improved match with the spectral-angular map will tend to significantly deteriorate the normal incidence match, and vice versa. Here we have attempted to minimize the degrees of freedom for more convenient fitting and to demonstrate the retrieval process.

The matching of the normal incidence data alone is straightforward. For TM polarization, one typically associates the permeability strictly with the longer wavelength (magnetic) resonance λ_m and the x- and z-permittivities with the shorter wavelength (electric) resonance. Incremental adjustments are made to the respective amplitudes, spectral positions, and sharpness of each function. The same process is done for TE polarization for the y-component of the permittivity. However, the difficulty that arises is that these functions must now also provide agreement with the spectral-angular data. Since only one component is responsible in the TE polarization, agreement with the spectral-angular map here occurs rather naturally. On the other hand, due to the TM polarization containing three functions that are responsible for the output, the agreement is not as straightforward. At this stage, further adjustments to these three functions are made such that the spectral-angular maps agree.

At first glance, the matching for the spectral-angular map is difficult to discern visually. To further clarify the matching success, in each map, we label each spot and characterize it by how much average relative intensity it receives. In this way, the three-dimensional plot can be reduced to a one-dimensional column graph, as shown in Figures 9 and 10. This is reasonable because only the material properties can provide the correct relative intensities, while geometry and diffraction dictate the angle and wavelength possible at each location. Since the material properties (i.e., permittivity and permeability) are responsible for the intensity of each spot, while the geometry and diffraction provide the spectral-angular location, this reduced one dimensional plot is best representative of the fitting, as this extracts only the desired effective optical parameters from the rest of the information producing Figures 7 and 8.

We can then compare intensities of each spot. Note that one set of permittivity and permeability must satisfy all sets of data, including normal incidence data as shown in Figure 5, which makes matching trustable.

As one can see from the comparative presentation in Figures 9 and 10, the agreement between the experiment and simulations are not ideal for some spots. For example, spots 3 and 4 in the TM 30° incident angle trial, which correspond to exit angles of approximately −65° and −45°, respectively, we note there is simulated radiation that is not quite detected experimentally. On the contrary, most all other spots at other incident angles are in close relative agreement. With slight deviations of the current optical properties' spectral position, amplitude, sharpness or symmetry, the "overall matching" drastically reduces. Here, overall matching simply translates to the matching of all eight spectral angular maps *and* the normal incidence transmission. For example, a 5-nm deviation in spectral position of the permeability may improve the matching for spots 3 and 4 in the TM 30° incident angle match, but subsequently worsen several other spots for other angles. Hence with the presented optical parameters (Figure 11), the best "overall match" was obtained.

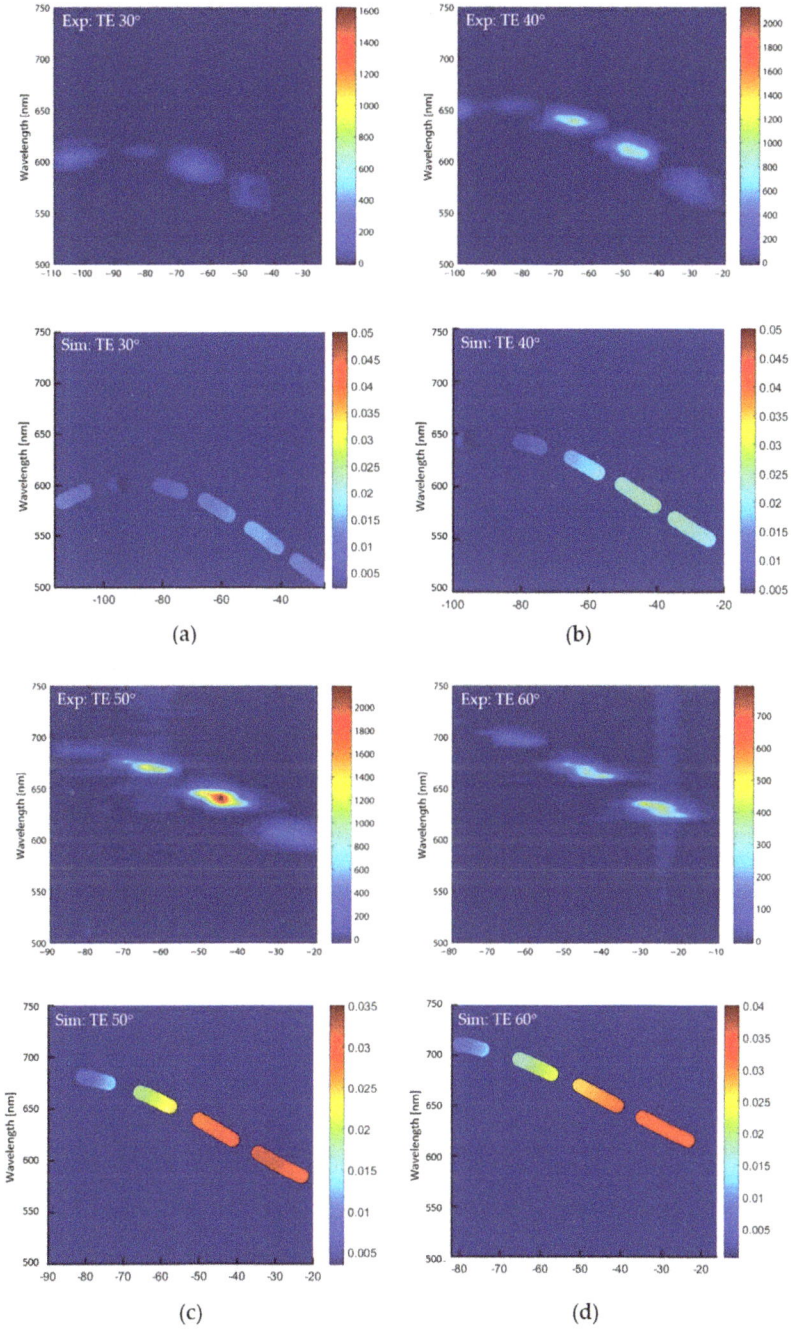

Figure 7. Experimental (top panels) and simulated (bottom panels) spectral-angular map of TE polarized incident light for incident angles 30° (**a**), 40° (**b**), 50° (**c**), and 60° (**d**). The intensity scales are in the same units for all maps of the TE polarization. Simulated data are matched to experimental data by considering maximum intensity in a spot.

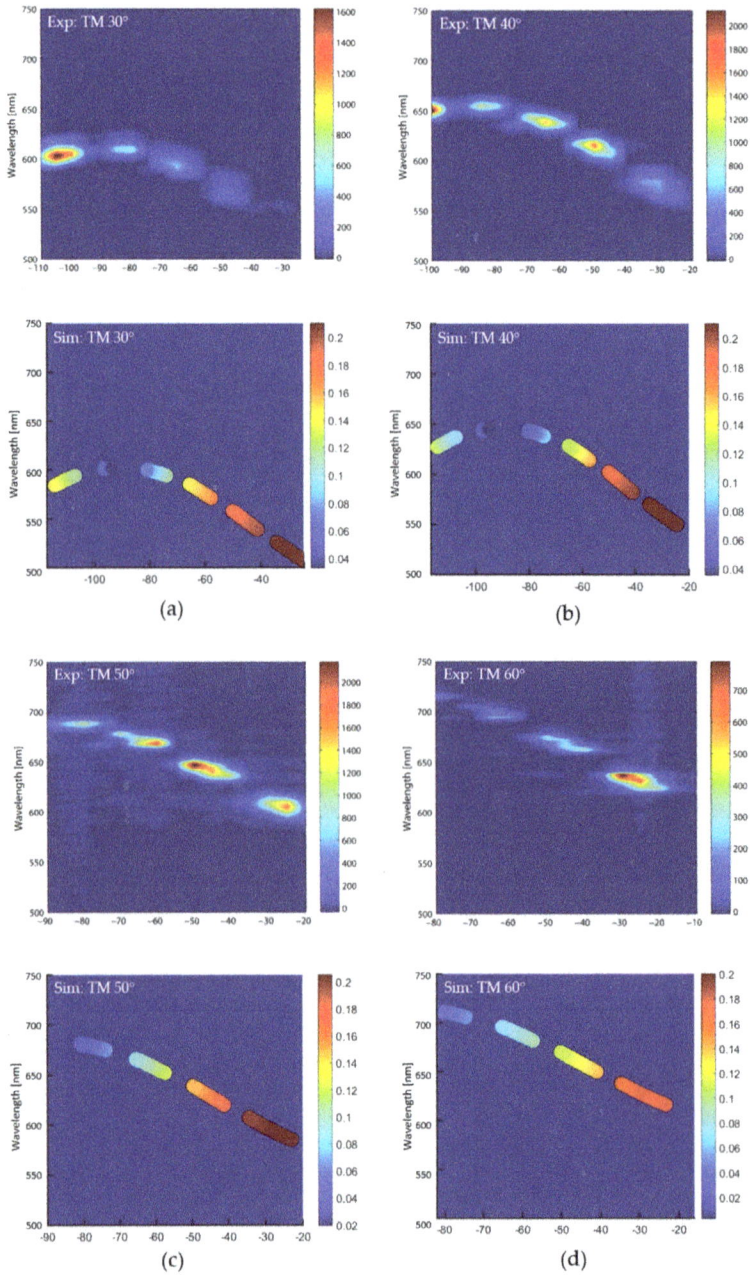

Figure 8. Experimental (top panels) and simulated (bottom panels) spectral-angular map of TM polarized incident light for incident angles 30° (**a**), 40° (**b**), 50° (**c**), and 60° (**d**). The intensity scales are in the same units for all maps of the TM polarization. Simulated data are matched to experimental data by considering maximum intensity in a spot.

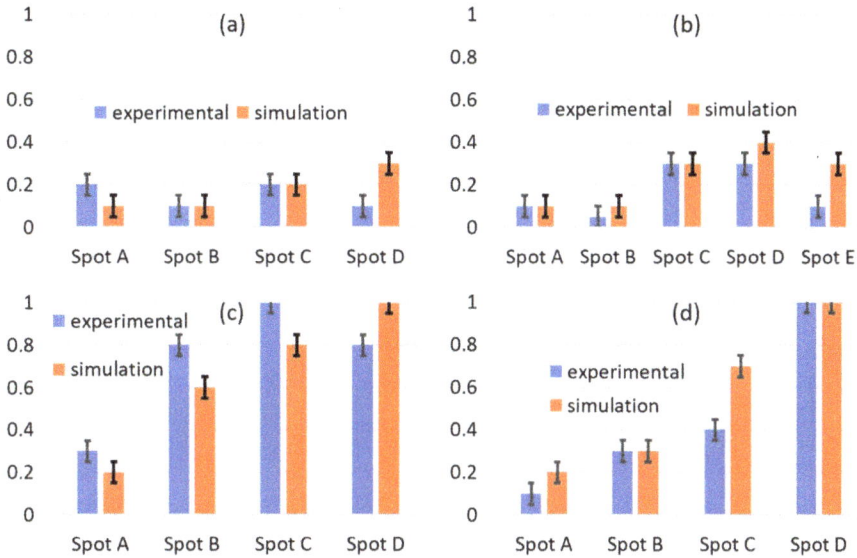

Figure 9. Spectral-angular data conversion for TE of each prominent "spot" to 1D column graph. The labels A, B, C, D, and E refer to the spots from left to right in Figure 7 in each intensity map at different incident angles: 30° (**a**), 40° (**b**), 50° (**c**), and 60° (**d**). Error bars reflect a 10% uncertainty. All intensities are on a relative zero to one scaling system.

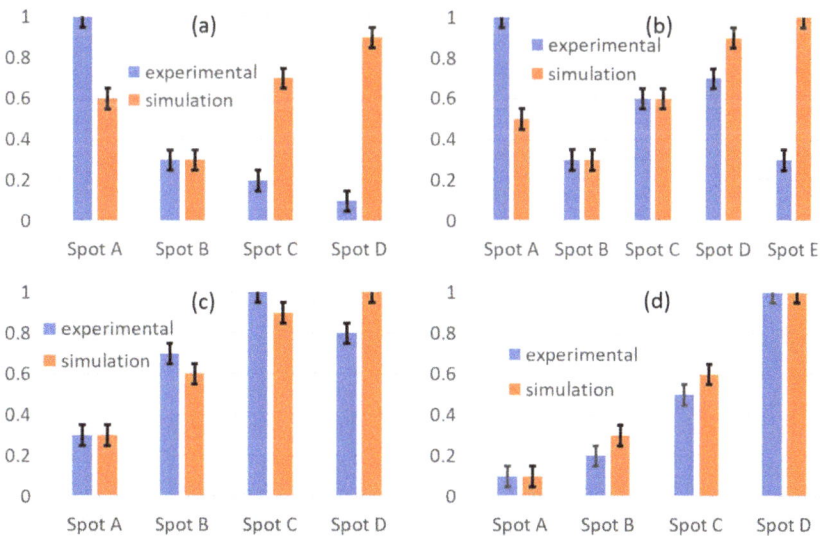

Figure 10. Spectral-angular data conversion for TM of each prominent "spot" to 1D column graph. The labels A, B, C, D, and E refer to the spots from left to right in Figure 8 in each intensity map, while panels a-d correspond to different incident angles 30° (**a**), 40° (**b**), 50° (**c**), and 60° (**d**). Error bars reflect a 10% uncertainty. All intensities are on a relative zero to one scaling system.

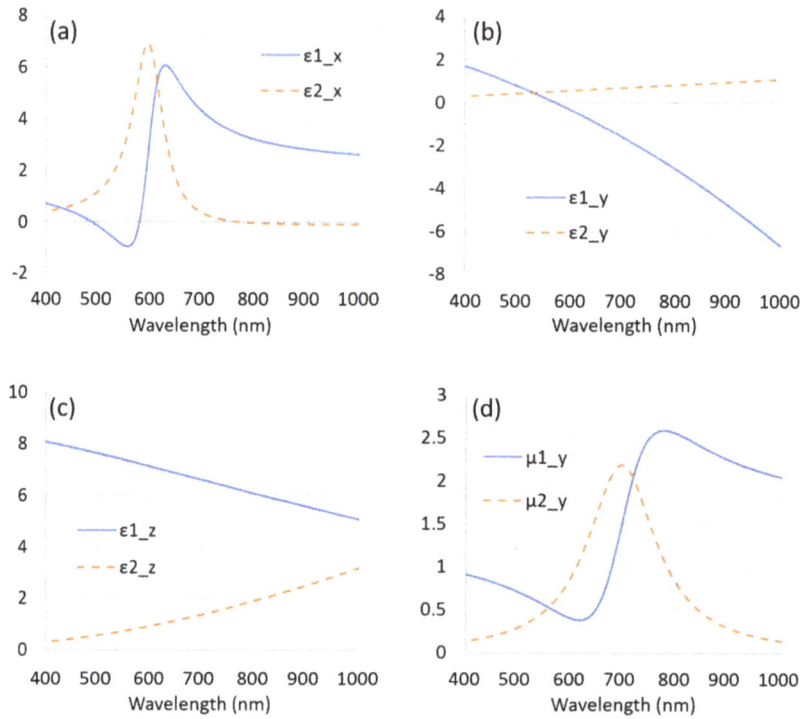

Figure 11. Retrieved parameters of the permittivity and permeability via our methods; note μ_x and μ_z (not pictured) are unity. (**a–c**) respectively show the x-, y-, and z-components of the permittivity while (**d**) shows the y-component of the permeability. ε_1 and μ_1 refer to real parts, while ε_2 and μ_2 refer to imaginary parts of the corresponding function.

Obtaining a good fit for all angles of incidence, in addition to normal incidence transmission gives all components of the permittivity and permeability of the effective strips. As always, the fitting is limited by the accuracy, to which we know the exact geometry, roughness, and any fluctuations that occur throughout the real grating structure.

The previous effective layer method works if normal incidence applications are required in which only the wavelengths longer than Wood anomaly are considered. This turns out normally to be reasonable for the visible spectrum. However, if one is trying to avoid the issue of diffraction in the wavelength range of interest, the period of the grating should be pushed to smaller dimensions, shifting the first diffraction event to shorter wavelengths. This begins to present a significant fabrication challenge. Even with a decreased period dimension, when oblique incidence is used with the metamaterial, the Wood anomaly is red shifted into the visible, creating an obvious problem for the effective layer method. In contrast, by using the effective strip retrieval, diffraction is accounted for, the period does not need to be pushed to smaller dimensions, and oblique incidence applications can be realized with the proper set of parameters. We introduce Table 1 below to summarize the comparison.

Note, that the main point of introducing magnetic response is that the two parameters, refractive index and impedance, become independent. If we can describe everything with just electric permittivity, thus $n = 1/Z$, and magnetic response is absent, meaning $\mu = 1$.

Table 1. Comparison between effective layer and strip methods for parameter retrieval.

Method	Advantages	Disadvantages	Limitations
Effective Layer	Well-known method; works well for most normal-incidence	Cannot be accurately used for oblique incidence	Affected by diffraction, only wavelengths longer than first Wood anomaly
Effective Strip	Accounts for diffraction; can be used for arbitrary angle of incidence	May require additional experimental setup	Long wavelength approximation

The capability of such a stacked grating material to waveguide the diffracted modes resulting from inclined illumination of the periodic grating surface makes it possible to apply such structures in biosensing. Using this design, the spectrum of the diffracted modes is sensitive to the refractive index of the material on the grating surface. For this reason, any changes in the refractive index due to a biochemical reaction on the surface [21] can be detected with this method. Additionally, upon retrieving the effective strip parameters, one may utilize such a grating design that exhibits a specialized set of optical properties in more practical situations where oblique incidence illumination is natural, such as the improvement of the efficiency in solar cells with unpolarized incident light [22]. Finally, we may generalize the method of parameter retrieval to other periodic nanostructures by using a similar experimental setup and simulation process. With a better understanding of how the effective optical properties of the strips depend on the geometry and materials chosen, it is hoped that further research will allow one to engineer a material with a specific set of optical parameters in mind which do not naturally occur.

4. Conclusions

We demonstrate a new approach for retrieving the effective optical properties of the structured strips of a metamaterial grating. The simulations are performed for an effective grating, implying uniform anisotropic material, using the well-known RCW technique, which is modified to allow magnetic behavior and anisotropy. Notably this expands the relevance of the model to capturing properties for inclined illumination by capturing the anisotropy of the properties of the strips in contrast to the validity at normal incidence only of the effective layer model. Coupling experimental measurements of samples with inclined illumination in addition to normal incidence to a scattering software tool (MIST) allows us to model both inclined and normal incidence illumination and capture relevant diffracted orders. Providing the proper set of complex permittivity and permeability functions, a successful fit to the experimental data will occur. This retrieval method allows for capturing behavior in nonzero diffraction orders. Indeed, the diffraction features observed in the experiment and simulations in Figure 5 at 450 nm have been excluded from the retrieval results in Figure 11 due to the applied approach. Thus our method provides a more broadly relevant effective property extraction for use in applications of magnetic gratings. The coupling of experimental methods to the MIST package is additionally useful in that, once the optical parameters are obtained, one may probe the system via the same simulation environment in other ways to realize applications, such as waveguide-based biosensing, and to optimize grating performance by examining configuration changes. We believe a similar scheme can be useful for two-dimensional gratings [23], i.e., the so-called "fishnet" nanostructure, to obtain their effective unit parameters.

Acknowledgments: Vladimir P. Drachev acknowledges support of this work by Russian Ministry of Education and Science grant RFMEFI58117X0026.

Author Contributions: Vladimir P. Drachev, Augustine M. Urbas and Dean P. Brown conceived and designed the experiments; David P. Lyvers, Kevin M. Roccapriore, Ekaterina Poutrina and Dean P. Brown performed experiments; Thomas A. Germer extended the RCW theory to include magnetic susceptibility, as described in Appendix A, Kevin M. Roccapriore and Thomas A. Germer performed numerical simulations. All authors contributed to overall data analysis and scientific discussions. Kevin M. Roccapriore, Vladimir P. Drachev,

David P. Lyvers, Augustine M. Urbas and Thomas A. Germer wrote the manuscript with contributions from all authors. Vladimir P. Drachev supervises the project.

Conflicts of Interest: The authors declare no conflicts of interest.

Appendix A

The RCW code used in this study implements the theory of Moharam et al. [15] as extended by Li [16] to properly account for the Fourier decomposition of the fields in the presence of discontinuities. To account for diagonal anisotropy and magnetic response of the media, the theory was further extended. For transverse electric (TE) polarization, the matrix in Equation (14) of Moharam's paper [15] in the presence of diagonal ε and μ is replaced by

$$\begin{bmatrix} 0 & \overline{M}_x \\ A & 0 \end{bmatrix}, \tag{A1}$$

where $A = K_x M_z^{-1} K_x - E_y$, and $\overline{M}_x$, M_z, and E_y are Toeplitz matrices formed from the Fourier coefficients of μ_x^{-1}, μ_z, and ε_y, respectively. V is then replaced by $V = \overline{M}_x^{-1} WQ$. Similarly, for transverse magnetic (TM) polarization, the matrix in Equation (34) of Moharam's paper [15] in the presence of diagonal ε and μ is replaced by

$$\begin{bmatrix} 0 & \overline{E}_x \\ B & 0 \end{bmatrix}, \tag{A2}$$

where $B = K_x E_z^{-1} K_x - M_y$, and $\overline{E}_x$, E_z, and M_y are Toeplitz matrices formed from the Fourier coefficients of ε_x^{-1}, ε_z, and μ_y, respectively. V is then replaced by $V = \overline{E}_x^{-1} WQ$.

References

1. Kildishev, A.V.; Cai, W.; Chettiar, U.K.; Yuan, H.-K.; Sarychev, A.K.; Drachev, V.P.; Shalaev, V.M. Negative refractive index in optics of metal-dielectric composites. *JOSA B* **2006**, *23*, 423–433. [CrossRef]
2. Brown, D.P.; Walker, M.A.; Urbas, A.M.; Kildishev, A.V.; Xiao, S.; Drachev, V.P. Direct measurement of group delay dispersion in metamagnetics for ultrafast pulse shaping. *Opt. Express* **2012**, *20*, 23082–23087. [CrossRef] [PubMed]
3. Drachev, V.P.; Podolskiy, V.A.; Kildishev, A.V. Hyperbolic metamaterials: New physics behind a classical problem. *Opt. Express* **2013**, *21*, 15048–15064. [CrossRef] [PubMed]
4. Ekinci, Y.; Christ, A.; Agio, M.; Martin, O.J.F.; Solak, H.H.; Löffler, J.F. Electric and magnetic resonances in arrays of coupled gold nanoparticle in-tandem pairs. *Opt. Express* **2008**, *16*, 13287–13295. [CrossRef] [PubMed]
5. Smith, D.R.; Schultz, S.; Markoš, P.; Soukoulis, C.M. Determination of effective permittivity and permeability of metamaterials from reflection and transmission coefficients. *Phys. Rev. B* **2002**, *65*, 195104. [CrossRef]
6. Smith, D.R.; Vier, D.C.; Koschny, T.; Soukoulis, C.M. Electromagnetic parameter retrieval from inhomogeneous metamaterials. *Phys. Rev. E* **2005**, *71*. [CrossRef] [PubMed]
7. Yuan, H.-K.; Chettiar, U.K.; Cai, W.; Kildishev, A.V.; Boltasseva, A.; Drachev, V.P.; Shalaev, V.M. A negative permeability material at red light. *Opt. Express* **2007**, *15*, 1076–1083. [CrossRef] [PubMed]
8. Kildishev, A.; Chettiar, U. Cascading optical negative index materials. *Appl. Comput. Electromagn. Soc. J.* **2007**, *22*, 172–183.
9. Nilsson, P.-O. Determination of Optical Constants from Intensity Measurements at Normal Incidence. *Appl. Opt.* **1968**, *7*, 435–442. [CrossRef] [PubMed]
10. Wood, R.W. Anomalous Diffraction Gratings. *Phys. Rev.* **1935**, *48*, 928–936. [CrossRef]
11. Hessel, A.; Oliner, A.A. A New Theory of Wood's Anomalies on Optical Gratings. *Appl. Opt.* **1965**, *4*, 1275–1297. [CrossRef]
12. Ni, X. PhotonicsSHA-2D: Modeling of Single-Period Multilayer Optical Gratings and Metamaterials. Available online: https://nanohub.org/resources/sha2d (accessed on 11 May 2017).

13. Germer, T.A. Modeled Integrated Scatter Tool (MIST). Available online: https://www.nist.gov/services-resources/software/modeled-integrated-scatter-tool-mist (accessed on 17 January 2018).

14. Moharam, M.G.; Gaylord, T.K. Rigorous coupled-wave analysis of grating diffraction—E-mode polarization and losses. *J. Opt. Soc. Am.* **1983**, *73*, 451–455. [CrossRef]

15. Moharam, M.G.; Grann, E.B.; Pommet, D.A.; Gaylord, T.K. Formulation for stable and efficient implementation of the rigorous coupled-wave analysis of binary gratings. *JOSA A* **1995**, *12*, 1068–1076. [CrossRef]

16. Li, L. Use of Fourier series in the analysis of discontinuous periodic structures. *JOSA A* **1996**, *13*, 1870–1876. [CrossRef]

17. Cai, W.; Shalaev, V. *Optical Metamaterials*; Springer: New York, NY, USA, 2010; ISBN 978-1-4419-1150-6.

18. Drachev, V.P.; Chettiar, U.K.; Kildishev, A.V.; Yuan, H.-K.; Cai, W.; Shalaev, V.M. The Ag dielectric function in plasmonic metamaterials. *Opt. Express* **2008**, *16*, 1186–1195. [CrossRef] [PubMed]

19. Schubert, M.; Tiwald, T.E.; Herzinger, C.M. Infrared dielectric anisotropy and phonon modes of sapphire. *Phys. Rev. B* **2000**, *61*, 8187–8201. [CrossRef]

20. Cai, W.; Chettiar, U.K.; Yuan, H.-K.; de Silva, V.C.; Kildishev, A.V.; Drachev, V.P.; Shalaev, V.M. Metamagnetics with rainbow colors. *Opt. Express* **2007**, *15*, 3333–3341. [CrossRef] [PubMed]

21. Liang, W.; Huang, Y.; Xu, Y.; Lee, R.K.; Yariv, A. Highly sensitive fiber Bragg grating refractive index sensors. *Appl. Phys. Lett.* **2005**, *86*, 151122. [CrossRef]

22. Kruk, S.S.; Wong, Z.J.; Pshenay-Severin, E.; O'Brien, K.; Neshev, D.N.; Kivshar, Y.S.; Zhang, X. Magnetic hyperbolic optical metamaterials. *Nat. Commun.* **2016**, *7*, 11329. [CrossRef] [PubMed]

23. Xiao, S.; Chettiar, U.K.; Kildishev, A.V.; Drachev, V.P.; Shalaev, V.M. Yellow-light negative-index metamaterials. *Opt. Lett.* **2009**, *34*, 3478–3480. [CrossRef] [PubMed]

MDPI

St. Alban-Anlage 66

4052 Basel

Switzerland

Tel. +41 61 683 77 34

Fax +41 61 302 89 18

www.mdpi.com

Applied Sciences Editorial Office

E-mail: applsci@mdpi.com

www.mdpi.com/journal/applsci

www.ingramcontent.com/pod-product-compliance
Lightning Source LLC
Chambersburg PA
CBHW041217220326
41597CB00033BA/6001